The Emerging WDM EPON

Mirjana Radivojević · Petar Matavulj

The Emerging WDM EPON

Mirjana Radivojević
Department of Computer Engineering
School of Computing Science, University
    Union
Belgrade
Serbia

Petar Matavulj
Department of Microelectronics
    and Engineering Physics
School of Electrical Engineering, University
    of Belgrade
Belgrade
Serbia

ISBN 978-3-319-85345-1   ISBN 978-3-319-54224-9   (eBook)
DOI 10.1007/978-3-319-54224-9

Jointly published with Academic Mind
ISBN: 978-86-7466-449-0 Academic Mind

Printed on acid-free paper

This Springer imprint is published by Springer Nature
The registered company is Springer International Publishing AG
The registered company address is: Gewerbestrasse 11, 6330 Cham, Switzerland

Dedicated to my parents, Milanka and Ratko Mirković, and to my family—my husband Goran and our children Dušan, Bogdan and Magdalena

—Mirjana Radivojević

Dedicated to my family

—Petar Matavulj

Preface

Motivation

Data transmission and networking technologies have witnessed tremendous growth over the past decade. However, much of this development and growth has been primarily in the core networks where high-capacity routers and ultra-high-capacity optical links have created a truly broadband infrastructure. The so-called first mile—the access network connecting end-users to backhaul infrastructure—remains a bottleneck in terms of the bandwidth and quality of service it affords to the end-users.

Until recently, the lack of broadband Internet access has been an inconvenience, but today this type of Internet access is truly essential for further development of many aspects of life. High-speed Internet expands our ability to communicate, learn, and entertain. Nowadays, Internet access is commonly associated with entertainment and information browsing, where most end-users use Internet connection for exchanging information, online gaming and applications such as Facebook and Twitter. On the other side, the number of business users has dramatically increased as network infrastructure is becoming the necessity for successful business operations. Service providers are mainly focused on the development and implementation of applications such as video on demand (VoD), high definition TV (HDTV), and online gaming which further increases the need for higher access speeds.

Moreover, broadband infrastructure is able to improve the quality of life in addition to supporting and improving business operations. Broadband online learning, video conferencing, online education for people with disabilities, national health network, and electronic health records are only a few examples that demonstrate the importance of the further development of broadband infrastructure.

In last decade, Ethernet passive optical networks (EPONs) were considered as a potential optimized architecture for the access network. An EPON is a point-to-multipoint fiber optical network designed to carry Ethernet frames at standard Ethernet rates. Propelled by rapid price declines of fiber optic and Ethernet components, these access networks combine the latest in optical and electronic technology and become the dominant players for delivering gigabit broadband connectivity to homes over a unified single platform. A major feature for this architecture is the use of a shared transmission media between all users; hence, medium access control arbitration mechanism is essential for the successful implementation of EPON. This mechanism ensures a contention-free transmission and provides end-users with an equal access to the shared media. Moreover, with the development of different multimedia-based applications such as HDTV, internet protocol TV (IPTV), and many others, the related quality of service (QoS) issues are also becoming a key concern. Namely, together with an increasing number of users in the access network, the number of bandwidth-hungry applications is increasing as well.

Until today, different models for incremental migration from TDM (Time Division Multiplexing) to TDM/WDM (Wavelength Division Multiplexing) EPON networks have been proposed, but most attention has been given to those solutions that support QoS implementation. Moreover, with the rapid development of different bandwidth applications, QoS support is becoming a key concern in WDM EPON network as it was the case with EPON networks.

Organization

Throughout the chapters of this book, we address many of the specified issues and present models and algorithms that we believe could resolve these issues. Moreover, we are convinced that the hardware implementation of the presented models and algorithms would operate in a foreseen manner and would not involve any additional implementation complexity. In the first place, we present a theory of differentiated services that essentially represents a basis for the QoS implementation in Ethernet-based networks since the QoS support has become a crucial requirement for a converged broadband access network with heterogeneous traffic. Furthermore, we present a detailed analysis and discuss the most important aspects of QoS support and implementation in networks that are based on optical transmission, particularly EPONs and WDM EPONs.

Following the outline of the theory of differentiated services, we present the overview of the EPON technology and introduce the readers with the challenges and unresolved issues in currently deployed EPON networks. We propose to use the multipoint control protocol (MPCP) defined within the IEEE (Institute of Electrical and Electronics Engineers) 802.3ah Task Force to arbitrate the transmission of different users, and we discuss the implementation of different dynamic bandwidth allocation (DBA) algorithms to effectively and fairly allocate bandwidths between end-users. In our first main contribution, we propose a novel dynamic intra-ONU (Optical Network Unit) scheduling algorithm for single-channel EPON systems, termed hybrid granting protocol (HGP), with full QoS support in accordance with the theory of differentiated services. For the presented model, we analyze how new scheduling algorithm can be combined with existing dynamic bandwidth allocation schemes in order to minimize packet delay and jitter. Specifically, the presented dynamic scheduling algorithm minimizes packet delay and jitter for delay and delay-variation sensitive traffic (e.g., voice transmissions) by allocating bandwidth in a GRANT-before-REPORT (GBR) fashion. This considerably improves their performance without degrading QoS guarantees for other service types. Detailed simulation experiments are presented for studying the performance and validating the effectiveness of the presented solutions.

However, given the steadily increasing number of users and bandwidth-hungry applications, although standardized, the current single-channel systems will not be able to satisfy the growing traffic demands in the future. Accordingly, we further suggest the implementation of wavelength division multiplexing technology in Ethernet passive optical networks in order to overcome problems that exist in the classical single-channel systems. Implementation of WDM technology would allow access network operators to respond to user requests for service upgrades and network evolution. Moreover, the deployment of WDM technology adds a new dimension to current TDM EPONs whereby the benefits of the new wavelength dimension are manifold. Among others, it may be exploited to increase network capacity, improve network scalability by accommodating more end- users, and separate services and service providers.

As previously discussed, the development of various multimedia applications requires QoS implementation in WDM EPONs just as it was the case with EPONs. In our next main contribution, we further present and analyze two models for wavelength and bandwidth allocation in the hybrid TDM/WDM EPON system with full QoS support, namely the fixed wavelength priority bandwidth allocation (FWPBA) model, and the dynamic wavelength priority bandwidth allocation (DWPBA) model. In order to implement QoS support, we present a new approach for QoS analysis and implementation in WDM EPON in which wavelength assignment takes place per service class and not per ONU, as suggested by the common approach in literature. In this way, the need for implementation of additional complex algorithms in order to support QoS, which increases system cost and increases the overall system efficiency, is avoided. Additionally, we propose a method for providing an upgrade to WDM in EPONs that includes an extension of the MPCP. In this way, the presented architecture and models will allow an incremental upgrade from TDM EPON to TDM/WDM EPONs. For each of the proposed models, wavelength and bandwidth allocation algorithms with full QoS support have been presented in order to fulfill all the requirements of new applications and services in a converged triple-play network. The presented models and algorithms are then compared in terms of average and maximum packet delay, packet variation delay (jitter), queue occupancy, packet loss rate, throughput, and overall system performance. Once again, detailed simulation experiments are presented for studying the performance and validating the effectiveness of the presented solutions.

In addition to FWPBA and DWPBA models, we present the future development of the proposed model that includes further segregation of medium priority traffic class according to the IEEE 802.1d standard and add another wavelength to be used for transmission of medium priority traffic, i.e., multimedia applications.

In the last chapter of the book, we conclude our work and present guidelines for future developments within this field.

Instruction for Readers

This book is intended to resolve misunderstood facts, equations, algorithms, and routines connected with EPONs, especially those directly connected to QoS and overall quality of traffic transmission in access networks. It may be very useful for students studying communication sciences, telecom operators, professionals in fields of optical networks and, of course, academic educators and scientists.

The first part of the book, namely Chapters I to IV, are intended to introduce readers to the field of EPON and describe and explain the most important facts about EPONs and the extension to WDM EPONs. These chapters recapitulate EPONs and QoS support in EPON and are more appropriate for students and readers that study EPONs for the first time. The second part of book, Chapters V to VII, are more advanced and introduce a novel concept of WDM EPON (wavelength assignment per service class) which may be designated as new generation EPON; a name which may seem more appropriate than "next generation EPON" which has been commonly used for WDM EPON and 10G EPON within the classical approach (wavelength assignment per ONU). These chapters are intended for readers with knowledge of basic architecture of PONs and EPONs, professionals in the field as well as for university educators at advanced levels of postgraduate studies (Master and Ph.D. courses).

We hope that the following text will be very useful for all readers and that it will open new frontiers for the future development of optical access networks.

Verba volant, scripta manent.

Belgrade, Serbia   Mirjana Radivojević
                                                                              Petar Matavulj

Contents

About the Authors

Dr. Mirjana Radivojević is an associate professor in the department for computer engineering at the School of Computing, University Union, Belgrade, Serbia. From 2008 to 2011, she was a teaching assistant, from 2011 to 2016, an assistant professor at the School of Computing, University Union, Belgrade, Serbia. From 2015, she also works as associate professor at the Basel School of Business, Basel, Switzerland.

Besides academic work, she has industrial experience working in Telco/ISP industries on design and implementation of multiservice networks. She has been involved in many telecom projects ranging from network/equipment operation and management to telecommunication planning, solution development, and design. Her research interests mainly include broadband access network, passive optical network, network modeling, and performance evaluation.

Dr. Petar Matavulj is professor of Physical Electronics at School of Electrical Engineering, University of Belgrade, Belgrade, Serbia. From 1994 to 2002, he was an assistant, from 2002 to 2008, an assistant professor, from 2008 to 2013, an associate professor, and from 2013, a full-time professor, at the School of Electrical Engineering, University of Belgrade. He is the author of more than 115 papers published in journals and conference proceedings. His research interests are modeling, simulation, and characterization of diverse optoelectronic devices, integrated photonics, and optical communications and networks.

Dr. Matavulj is a member of the IEEE Photonics Society (PS), Electron Device Society (EDS), and Communications Society (ComSoc), OSA (Optical Society of America) and ODS (Optičko Društvo Srbije —Optical Society of Serbia).

Acronyms

3GPP EVDO	3rd Generation Partnership Project 2 EVDO
3GPP LTE	3rd Generation Partnership Project LTE
3play	Triple-play network
ADSL	Asymmetric DSL
ADSL2+	Asymmetric DSL+
AF	Assured Forwarding
AMPS	Advanced Mobile Phone System
AON	Active Optical Network
APD	Avalanche PhotoDiode
APON	ATM PON
ATM	Asynchronous Transfer Mode
AWG	Arrayed Waveguide Grating
BA	Behavior Aggregate
BE	Best Effort
BPL	Broadband over Power Line
BPON	Broadband PON
BS	Band Splitter
CATV	Community Antenna TeleVision
CDMA	Code Division Multiple Access
CDMA EVDO	CDMA EVolution-Data Optimized
CDR	Clock and Data Recovery
CMTS	Cable Modem Termination System
COPS	Common Open Policy Service
CPON	Composite PON
CQ	Custom Queuing
CRC	Cyclic Redundancy Check
CS	Class Selector
CSMA/CD	Carrier Sense Multiple Access with Collision Detection
CWDM	Coarse WDM
DA	Destination Address

DBA	Dynamic Bandwidth Allocation
DBA-CL	DBA CycLe
DBRu	Dynamic Bandwidth Report
DEMUX	DEMUltipleXer
DOCSIS	Data Over Cable modem Service Interface Specification
DP	Drop Probability
DS	Differentiated Services
DSCP	Differentiated Services Code Point
DSL	Digital Subscriber Line
DSLAM	DSL Access Modules
DWBA	Dynamic Wavelength and Bandwidth Allocation
DWDM	Dense WDM
DWDT	Dynamic Wavelength Dynamic Time
DWPBA	Dynamic Wavelength Priority Bandwidth Allocation
DWPBA-FS	DWPBA with Fine Scheduling
ECN	Explicit Congestion Notification
EDGE	Enhanced Data rate for Global Evolution
EF	Expedited Forwarding
EPON	Ethernet based PON
FEC	Forward Error Correction
FIFO	First In, First Out
FSAN	Full Service Access Network
FTTB	Fiber To The Building
FTTC	Fiber To The Curb
FTTH	Fiber To The Home
FTTN	Fiber To The Node
FTTP	Fiber To The Premises
FTTx	Fiber To The x (x stands for node)
FWPBA	Fixed Wavelength Priority Bandwidth Allocation
GAR	GRANT after REPORT
GBR	GRANT before REPORT
GEM	GPON Encapsulation Method
GEO	GEostationary (Earth) Orbit
GFP	Generic Frame Procedure
GMII	Gigabit Media Independent Interface
GPON	Gigabit PON
GPRS	General Packet Radio Service
GSM	Global System for Mobile communication
GTC	GPON Transmission Convergence
GTG	GATE-To-GATE delay
GTR	GATE-To-REPORT delay
HDTV	High-Definition TeleVision
HG	Hybrid Grant protocol
HG-PBS	Hybrid Granting algorithm with Priority-Based Scheduling
HSDPA	High Speed Downlink Packet Access

IETF	Internet Engineering Task Force
IP	Internet Protocol
IPACT	Interleaved Polling with Adaptive Cycle Time
IPL	Internet over Power Line
ISDN	Integrated Services Digital Network
ISP	Internet Service Providers
ITU-T	International Telecommunication Union-Standardization sector
JIT	Just-In-Time scheduling
LAN	Local Area Network
LARNET	Local Access Router NETwork
LEO	Low Earth Orbit
LLC	Logical Link Control
LLID	Logical Link Identification
LOS	Line Of Sight
LTE	Long Term Evolution advanced
MAC	Medium Access Control
MDI	Medium Dependent Interface
MDU/MTU	Multiple Dweling Unit/Multiple Tenant Unit
MEO	Medium Earth Orbit
MFL	Multi Frequency Laser
MG-IPACT	Modified Gated IPACT
MIB	Management Information Base
MPCP	Multi-Point Control Protocol
MPCP-CL	MPCP CycLe
MPCPDU	MPCP Data Units
MTTR	Mean Time To Repair
MTU	Maximum Transmission Unit
MUX	MUltipleXer
NGN	Next Generation Network
NID	Network Interface Device
NMT	Nordic Mobile Telephone
OAM	Operations, Administration, and Maintenance
OAMP	Operation, Administration, Maintenance and Provisioning
ODN	Optical Distribution Network
OECD	Organization for Economic Co-operation and Development
OLT	Optical Line Terminal
OMCC	ONU Management and Control Channel
OMCI	ONT Management and Control Interface
ONT	Optical Network Terminal
ONU	Optical Network Unit
OSI	Open Systems Interconnection
P2MP	Point-to-MultiPoint
P2P	Peer-to-Peer
PBS	Priority Based Scheduling

PCBd	Physical Control Block Downstream
PCS	Physical Coding Sublayer
PD	PhotoDetector or PhotoDiode
PDC	Personal Digital Cellular
PDP	Policy Decision Point
PEP	Policy Enforcement Point
PHB	Per-Hop Behavior
PIN	Positive Intrinsic Negative
PLC	Power Line Communication
PLOAM	Physical Layer Operation, Administration, and Maintenance
PLOu	Physical Layer Overhead
PLT	Power Line Telecommunication
PMA	Physical Medium Attachment sublayer
PMD	Physical Media Dependent sublayer
PN	Passive Node
PON	Passive Optical Network
POTS	Plain Old Telephone Service
PPP	Point-to-Point Protocol
QoS	Quality of Service
RAP	Resource Allocation Protocol
READSL2	Reach Extended ADSL2
RITENET	Remote Interrogation of TErminal NETwork
RN	Remote Node
RS	Reconciliation Sublayer
RSVP	Resource reSerVation Protocol
RTG	REPORT-To-GATE (delay)
RTS	REPORT-To-Schedule (delay)
RTT	Round-Trip Time
SA	Source Address
SAR	Segmentation And Reassembly
SBS	Strict-Based Scheduling
SDH	Synchronous Digital Hierarchy
SDSL	Symmetric DSL
SLD	Start of the LLID Delimiter
SME	Shared-Medium Emulation
SONET	Synchronous Optical Network
SPON	(DWDM) Super PON
STG	Schedule-To-GATE (delay)
SUCCESS	Stanford University aCCESS architecture
SUCCESS-DWA	SUCCESS Dynamic Wavelength Allocation
SUCCESS-HPON	SUCCESS Hybrid PON
SWDT	Static Wavelength Dynamic Time
TACS	Total Access Communication System
TCP	Transmission Control Protocol
TDM	Time Division Multiplexed

TDMA	Time Division Multiple Access
TD-SCDMA	Time Division Synchronous Code Division Multiple Access
ToS	Type of Service
UMB	Ultra Mobile Broadband
UMTS-HSDPA	Universal Mobile Terrestrial System with HSDPA
VC	Virtual Circuit
VCI	Virtual Circuit Identification
VDSL	Very High Bit Rate DSL
VoD	Video-on-Demand
VoIP	Voice over IP
VP	Virtual Path
VPI	Virtual Path Identification
VPN	Virtual Private Network
WCDMA	Wideband CDMA
WDM	Wavelength Division Multiplexing
WDMA	Wavelength Division Multiple Access
WDM IPACT-ST	WDM IPACT with a Single polling Table
WFQ	Weight Fair Queuing
WiMAX	Worldwide interoperability for Microwave Access
WMAN	Wireless Metropolitan Area Network
WRR	Weighted Round Robin
WWAN	Wireless Wide Area Network

Chapter 1
Introduction

1.1 Traffic Growth and First Mile Evolution

Broadband Internet access (or just 'broadband') is a high data rate connection to the Internet that is typically contrasted with the dial-up access using a 56 kbps modem. Dial-up modems are limited to a bit rate of about 60 kbps and require the dedicated use of a telephone line, whereas broadband technologies supply more than this rate generally without disrupting telephone use. Broadband is a term that is used consistently with different types of Internet connections. Broadband in telecommunications means a wide range of frequencies that are available to transmit information. This eventually means that the wider the range of frequencies available, the higher the amount of information that can be sent at any given point of time will be. Since data transmission and networking technologies have witnessed tremendous growth over the past decade, the number of users and bandwidth-hungry applications has significantly increased as well.

However, much of this development and growth has been primarily in the core networks where high capacity routers and ultrahigh-capacity optical links have created a truly broadband infrastructure. Namely, with the expansion of services offered over the Internet, a dramatic increase in bandwidth has been facilitated in the backbone network through the use of wavelength-division multiplexing (WDM), providing tens of gigabits per second per wavelength. Moreover, a wide range of increasingly bandwidth-intensive services are continuing to emerge, e.g., storage extension/virtualization, grid computing, and packet video teleconferencing. At the same time, end-user local area networks (LANs) have also seen their tributary speeds progressively increase from 100 Mbps upward to 1 Gbps and beyond. Such a growing gap between the capacity of the backbone network and end-users' needs results in a serious bottleneck in the access network between them [1], i.e., 'access bottleneck,' as in Fig. 1.1. The term 'last mile' used in the literature to describe the access network is now often replaced by the 'first mile' in order to

© Academic Mind and Springer International Publishing AG 2017   1
M. Radivojević and P. Matavulj, *The Emerging WDM EPON*,
DOI 10.1007/978-3-319-54224-9_1

emphasize the role and importance of this network segment for the further development of information-communication networks.

Today, the term 'residential broadband' describes the group of technologies that provide high-bandwidth connection to the Internet for residential consumers. The replacement for the now fading residential dial-up technology should allow end-users to use different services such as watching video stream, downloading music in seconds, video and voice chats, and real-time gaming, and many others.

Broadband is often called high-speed Internet access, because it usually has a high rate of data transmission. As shown in Fig. 1.2, average broadband speed can go up to 60 Mbps. In general, any connection to the customer of 256 kbps or greater is more concisely considered broadband Internet access. The ITU-T

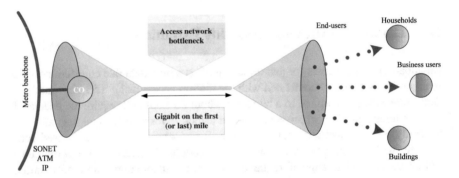

Fig. 1.1 Access network bottleneck

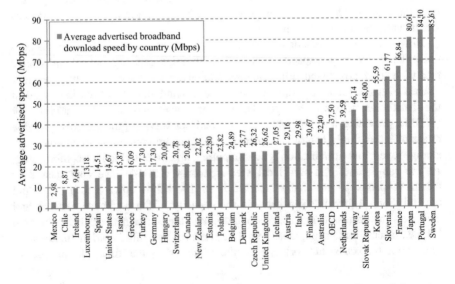

Fig. 1.2 Average advertised broadband download speed by country. *Source* Information technology and innovation foundation, September 2010

(International Telecommunication Union Standardization Sector) recommendation I.113 has defined broadband as a transmission capacity that is faster than primary rate ISDN, at 1.5–2 Mbps. The US Federal Communications Commission definition of broadband amounts to 4.0 Mbps. The OECD (Organization for Economic Cooperation and Development) has defined broadband as 256 kbps in at least one direction, and this bit rate is the most common baseline that is marketed as 'broadband' around the world. Many countries have already projected the increase in broadband access speed and penetration. USA expects that every household should have the access speed of 4 Mbps by 2020, UK expects to provide everyone with at least 2 Mbps by 2015, Korea expects to raise average speeds to 10 Mbps by 2012, EU to provide at least 30 Mbps for every EU citizens by 2020, Russia to have 35% broadband penetration by 2015, and Serbia to have 20% broadband penetration with 4 Mbps by 2012 [2, 3].

In practice, the advertised maximum bandwidth is not always reliably available to the customer. Namely, physical link quality can vary, and Internet service providers (ISPs) usually allow a greater number of subscribers than their backbone connection or neighborhood access network can handle, under the assumption that most users will not be using their full connection capacity very frequently. This aggregation strategy known as a contended service works more often than not, so users can typically burst to their full bandwidth most of the time. However, applications such as peer-to-peer (P2P) file sharing systems often require extended durations of high-bandwidth usage and consequently violate these assumptions and most probably cause major problems for ISPs.

In general, broadband solutions can be classified into two groups: fixed line technologies and wireless technologies. The fixed line solutions communicate via a physical network that provides a direct wired connection from the customer to the service supplier. Wireless solutions use radio or microwave frequencies to provide a connection between the customer and ISP network.

Fixed line broadband technologies rely on direct physical connection between subscribers and service suppliers. Many broadband technologies such as cable modem, digital subscriber line (DSL) technologies. and broadband over power line (BPL) have evolved to use an existing form of subscriber connection as the medium for communication. Namely, xDSL systems use the twisted copper pair traditionally used for voice services like plain old telephone service (POTS). BPL technology uses the power lines installed in subscriber homes to carry broadband signals. On the other hand, cable modems use existing hybrid fiber–coax (HFC) cable TV networks. When the above-mentioned technologies are not available, satellite connections can be used instead.

For example, Fig. 1.3 shows the growth and penetration of residential broadband technologies including DSL, BPL, and cable technologies for the US market in the last eight years. The development of broadband market is characterized as follows:

- fast growth, 11% increase per year;
- more than 50% penetration already reached across Internet households;

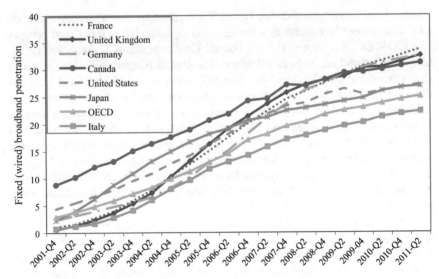

Fig. 1.3 Fixed (wired) broadband penetration, G7 countries. *Source* OECD June 2011 www. oecd.org/sti/ict/broadband

- several competing broadband service providers; and
- telephone companies, wireless carriers, cable TV service providers, and satellite providers.

In addition to the stated technology, a lot of attention nowadays is given to optical networks as a possible solution for the bottleneck problem in the access network. Namely, none of the aforementioned technologies, including xDSL, BPL, HFC, and wireless technologies, are able to offer sufficient bandwidth for the successful transmission of different video and multimedia applications. Consequently, a lot of attention in the Telco industry and market is given to the FTTx technologies as the main candidate for the successful realization of a truly broadband infrastructure.

Fiber to the x (FTTx) is a generic term for any broadband network architecture using optical fiber to replace all or part of the usual copper local loop used for last mile, as in Fig. 1.4. Today, the FTTx term includes several configurations of fiber deployment, where x stands for node (FTTN), premises (FTTP), building (FTTB), curb (FTTC), and home (FTTH).

These solutions require the installation of new fiber (link) from the local exchange (central office) directly to or closer to the subscriber. Fiber installation can offer the ultimate in broadband bandwidth capability, but installation cost of such network must be taken into account.

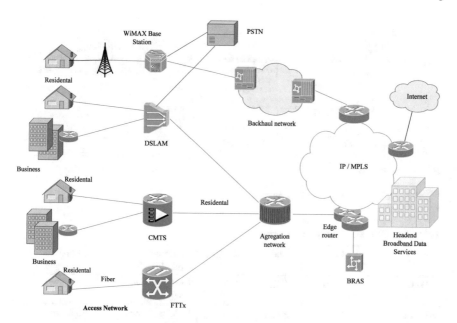

Fig. 1.4 Broadband network

1.1.1 DSL Technologies

Digital subscriber line (DSL) is a family of technologies that provides digital data transmission over the wires of a local telephone network. DSL originally stood for 'digital subscriber loop'. DSL service is delivered simultaneously with the regular telephone service on the same telephone line, i.e., using the existing copper telephone infrastructure to facilitate high-speed data connections. When using a modem on a regular telephone line, the line is busy and cannot be used to make or receive a phone call at the same time. With DSL technology, the phone line can carry two signals at the same time (Fig. 1.5):

- a phone call/fax/analog-modem connection;
- a high-speed digital signal for Internet access.

DSL achieves this by dividing the voice and data signals on the telephone line into three distinct frequency bands. These frequency bands are subsequently separated by filtering on the customer side. On the service provider's side, DSL access modules (DSLAMs) are placed in the local exchange or at nodes in the access network in order to transmit and receive data signals.

Today, there are various DSL technology options as shown in Table 1.1 [4–9]. In practice, actual speeds may be reduced depending on line quality where the most significant factor in line quality lies in the distance from the DSLAM to the customer's equipment. The key technologies are asymmetric DSL (ADSL), symmetric

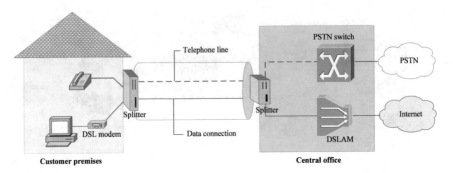

Fig. 1.5 DSL technology implementation

Table 1.1 DSL technology characteristics

Family	ITU	Name	Ratified	Maximum speed capabilities[a]	Maximum distance
ADSL	G.992.1	G.dmt	1999	8 Mbps down, 1 Mbps up	5.4 km
ADSL2	G.992.3	G.dmt.bis	2009	12 Mbps down, 1.4 Mbps up ITU G.992.3 annex M: 3.3 Mbps up	5.6 km
ADSL2+	G.992.5	G.adslplus	2009	24 Mbps down, 1 Mbps up	1.5 km
G.SHDSL	G.991.2	G.shdsl	2003	2.3 Mbps down, 2.3 Mbps up (two wires) 4.6 Mbps down/4.6 Mbps up (four wires)	Up to 6 km (two wires) Up to 5 km (four wires)
VDSL	G.993.1	G.vdsl	2004	50 Mbps down, 6.4 Mbps up	1.5 km
VDSL2–12 MHz long reach	G.993.2	G.vdsl2	2006	50 Mbps down, 30 Mbps up	1.0 km
VDSL2–30 MHz short reach	G.993.2	G.vdsl2	2006	100 Mbps down/30 Mbps up	0.3 km

[a]At maximum distance, the achieved rates are significantly smaller than the maximum rates defined in the fifth column; actual speeds may vary based on factors such as line quality, distance from exchange (for ADSL/ADSL2+), technology used, hardware capabilities, server route, and network congestion

DSL (SDSL), very high bit-rate DSL (VDSL), and ADSL2+. The data throughput of consumer DSL services typically ranges from 256 kbps to 100 Mbps in the direction to the customer (downstream), depending on the DSL technology, line conditions, and service-level implementation.

DSL is currently the most prevailing broadband choice in the world with over 65% market share and more than 200 million users. DSL is available in every region of the world, and ADSL owns the majority of the market, even though VDSL and ADSL2+ are gaining ground. However, DSL is a distance-sensitive technology where the signal quality decreases and the connection speed goes down with the increase in the distance between the DSLAM and users, as in Figs. 1.6 and 1.7.

Since the most popular DSL technologies including ADSL, ADSL2, and ADSL2+ have not solved the problem of bottlenecks, they are not able to offer sufficient amount of bandwidth for the successful transmission of different multimedia applications, as in Fig. 1.7. This resulted in VDSL and VDSL2 technologies gaining a lot of attention as these technologies are seen as a possible key for enabling real competition between Telco's and cable operators.

Table 1.1 shows that ADSL technology can provide maximum downstream speeds of up to 8 Mbps and upstream speeds of up to 1 Mbps. The maximum distance for ADSL service is 5.4 km, but at this distance, transmission speeds are limited to approximately 500 kbps. For different business applications, customers could use GSHDSL which allows high-speed download and upload, but again the maximum available bandwidth is approximately 3 Mbps. With, for example, video-on-demand (VoD) service requiring at least 3 Mbps and high-definition television (HDTV) requiring approximately 15–20 Mbps, it is clear that neither ADSL nor GSHDSL can meet the bandwidth requirements for HDTV or even basic video service over the full network. However, VDSL and ADSL2+ can offer enough bandwidth to allow video services. VDSL can offer up to 52 Mbps, but

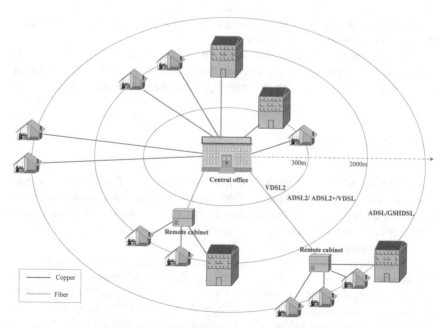

Fig. 1.6 DSL technologies—distance versus speed

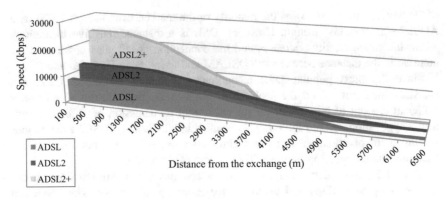

Fig. 1.7 Characteristics of various ADSL technologies

only at very short distances. In order to offer VDSL to the significant proportion of customers, the DSLAM needs to be relocated to street cabinets, closer to the subscriber (fiber feeds are now installed in the street cabinets). The cost of this upgrade is equal to the cost of laying fiber to cabinets; i.e., VDSL is prohibitively expensive relative to ADSL, and therefore, VDSL deployments are limited.

The latest technologies to emerge from the DSL family are ADSL2+ and ADSL2++. ADSL2++ is not yet supported with the appropriate standard. ADSL2+ is standardized and allows transmission of sufficient bandwidth for some video services over greater distances than VDSL without the need for DSLAM reallocation. As a result, ADSL2+ is becoming the upgrade path for operators wishing to improve their standard ADSL offerings.

DSL technologies are currently dominant in Europe, as in Fig. 1.8. Research conducted by European Cable Association during 2010 shows that different variants of DSL technology are used by the majority of the population in European Union.

1.1.2 Cable Technology

The abbreviation CATV is often used to mean 'cable TV'. Back in 1984, it origarinally stood for Community Antenna Television in areas where signal reception was limited by the distance from transmitters or mountainous terrain. To overcome the stated problem, large 'community antennas' were constructed, and cable was run from them to individual homes.

The traditional CATV network is an all coaxial cable network, as in Fig. 1.9. The headend receives the TV signal typically from large satellite dishes, microwave, or fiber-optic feeds. The received TV signal is then broadcast to the consumer via a tree and branch coaxial cable architecture where CATV cables are predominately aerial and buried cables. The described traditional coaxial architecture is a

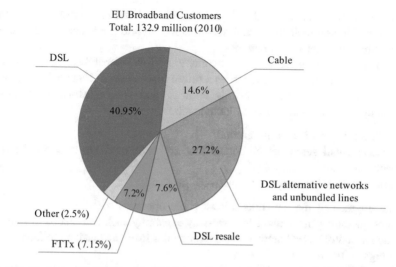

EU Broadband Customers
Total: 132.9 million (2010)

DSL

Cable

14.6%

40.95%

27.2%

DSL alternative networks
and unbundled lines

Other (2.5%)

7.2% 7.6%

FTTx (7.15%)

DSL resale

Fig. 1.8 Deployment of DSL technologies in the European Union. *Source* European Cable Association, http://www.cableeurope.eu/index.php?page=ff-2009-broadband-in-europe

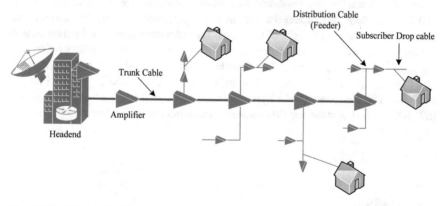

Distribution Cable
(Feeder)

Subscriber Drop cable

Trunk Cable

Amplifier

Headend

Fig. 1.9 Traditional CATV network

one-way network; i.e., it cannot return any data from the end-users back to the network headend. Consequently, this network is unsuitable for voice and data communication services.

As shown in Fig. 1.9, trunk cables carry TV signal from the headend to branch points where the signal is amplified and routed to either feeder cables or directly to distribution cables. Additionally, trunk cables may carry the feed to nodes where it is then distributed. Feeder cables receive the signal from trunk cables and carry the signal deeper into the community where it is transferred to the distribution cables and further to end-users. As explained above, the traditional CATV network restricted its use to one-way broadcasting of TV and video signals. As CATV companies began expanding into telephony, data, and Internet access services, they began transitioning their traditional networks to a hybrid fiber/coaxial architecture, commonly called the HFC network.

In order to upgrade traditional networks to two-way capability, operators decided to invest in fiber, which gave them a future proof network that could ultimately deliver almost unlimited bandwidth both to and from customers' premises. Upgrading coax networks to HFC includes the following:

- putting a lot of fiber in the ground;
- creating small segments of homes that are connected via coax to the fiber backbone; and
- installing new hardware at the network headends.

The upgrade of the networks allows cable operators to offer a new range of services, including those requiring two-way capability such as broadband Internet, telephony, or VoD. HFC networks use fiber cables from the headend to feed nodes, as in Figs. 1.10 and 1.11.

The HFC network can be operated bidirectionally, meaning that signals are carried in both directions on the same network from the headend/hub office to the home and from the home to the headend/hub office. The downstream signals carry information from the headend/hub office to the home, such as video content, voice, and Internet data. The upstream signals carry information from the home to the headend/hub office, such as control signals for ordering a movie or Internet data or for sending e-mails. The downstream and upstream paths are actually carried over the same coaxial cable in both directions between the optical node and the home. In order to prevent the interference of signals, the frequency band is divided into two sections: The first one (5–42 MHz) is used for upstream, and the second one (52–1000 MHz) is used for downstream communications. Different countries use

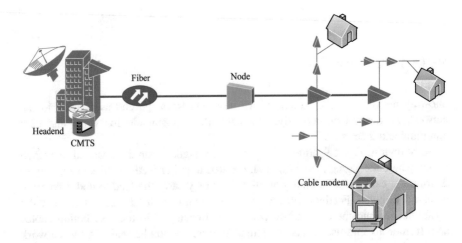

Fig. 1.10 HFC network

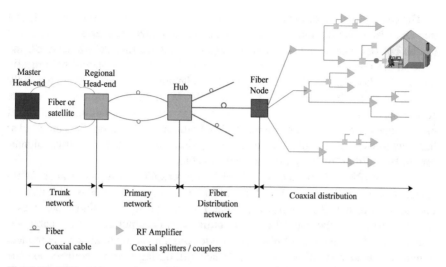

Fig. 1.11   HFC architecture

different band sizes, but are similar in that there is much more bandwidth for downstream communication compared to upstream communication.

The fiber-optic network extends from the cable operators' master headend, sometimes to regional headends, and out to a neighborhood's site, and finally to a fiber-optic node which serves anywhere from 25 to 2000 homes, as in Fig. 1.11. A master headend will usually have satellite dishes for reception of distant video signals as well as IP aggregation routers. Some master headends also house telephony equipment for providing telecommunications services to the community. A regional or area headend/hub will receive the video signal from the master headend where various services are encoded, modulated, and converted onto RF carriers, combined onto a single electrical signal and inserted into a broadband optical transmitter. This optical transmitter converts the electrical signal to a downstream optically modulated signal that is sent to the nodes. Fiber-optic cables connect the headend or hub to optical nodes in a point-to-point or star topology. The node serves as an equipment location and the interface point with the distribution network. In the node, the optical signal is converted to an electrical signal for transmission to customers via ordinary coaxial cables.

The optical portion of the network provides a large amount of flexibility. If there are not many fiber-optic cables to the node, wavelength-division multiplexing can be utilized to combine multiple optical signals onto the same fiber. Optical filters are used to combine and split optical wavelengths onto the single fiber. For example, the downstream signal could be on a wavelength at 1550 nm, and the return signal could be on a wavelength at 1310 nm. There are also techniques to put multiple downstream and upstream signals on a single fiber by putting them at different wavelengths.

The coaxial portion of the network connects 25–2000 homes (500 is typical) in a tree-and-branch configuration from the node. To minimize the number of repeaters on the coaxial cables and to provide telephony and Internet access services, the node sizes must be kept relatively small. In practice, node size should be typically under 1000 households (in the traditional architecture, node sizes averaged 2000 homes, but could be many times that size) [10]. With the development of Internet access services, the node size must be further reduced to minimize electrical interference on the remaining coaxial cables. Radio frequency amplifiers are used at intervals to overcome cable attenuation and passive losses of the electrical signals caused by splitting or 'tapping' the coaxial cable.

At a subscriber's home, a coax splitter is used to split the drop coaxial cableinto several coaxial cables, as in Fig. 1.12. Most frequently, one cable is allocated for Internet connection and the rest for cable TV service. Internet signal occupies a separate channel in the coaxial cable spectrum so it does not disturb the cable TV channels. A cable modem at a customer's side communicates with the cable modem termination system (CMTS) over CATV network during Internet connection. The communication follows DOCSIS (Data over Cable Modem Service Interface Specification) standard [11]. CMTS at the headend terminates connections from all end-users cable modems. It is very important to emphasize that Internet access over cable TV network is delivered through a medium (coax cable) which is shared by all subscribers within the coverage of a neighborhood hub.

The maximum shared data rate is 42 Mbps downstream and 30 Mbps upstream. Real-world performance depends on how many people within a neighborhood connect to the Internet at the same time, the connection to backbone network, and Internet traffic load. However, most operators offer minimum downstream and upstream rates during peak hours.

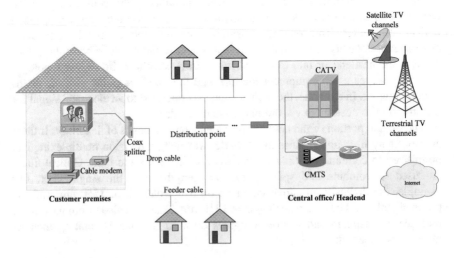

Fig. 1.12 HFC subscribers

Table 1.2 Maximum synchronization speed (maximum usable speed)

Version	DOCSIS		EuroDOCSIS	
	Downstream	Upstream	Downstream	Upstream
1.x	42.88 (38) Mbps	10.24 (9) Mbps	55.62 (50) Mbps	10.24 (9) Mbps
2	42.88 (38) Mbps	30.72 (27) Mbps	55.62 (50) Mbps	30.72 (27) Mbps
3.0, 4 channel	171.52 (152) Mbps	122.88 (108) Mbps	222.48 (200) Mbps	122.88 (108)Mbps
3.0, 8 channel	+171.52 (+152) Mbps	122.88 (108) Mbps	444.96 (400) Mbps	122.88 (108) Mbps

Currently, a vast majority of cable operators around the world are focused on the implementation of the newest DOCSIS standard, called DOCSIS 3.0, as in Table 1.2. This standard was released in 2006 and addresses the following most important service goals of next-generation HFC networks [11]:

- increasing channel capacity;
- enhancing network security;
- expanding addressability of network elements; and
- deploying new service offerings.

Stated services are supported through various techniques and features including the following:

- channel bonding (upstream and downstream channel bonding);
- support for IP multicast and features such as source-specific multicast (SSM) and QoS (quality of service) support for multicast;
- QoS support and authorization;
- support for IPv6 (IPv6 provisioning and management of cable modems);
- enhanced security;
- physical layer enhancements including upstream frequency range extension, and topology and ambiguity resolution; and
- enhanced diagnostic, management, and monitoring.

Cable Internet and services are most commonplace in North America, Europe, Australia, and East Asia, though they are also present in many other countries, mainly in South America and the Middle East. Cable TV has had little success in Africa, as it is not cost-effective to lay cables in sparsely populated areas.

1.1.3 BPL (Broadband Power Line) Technology

BPL technology, also known as broadband over power lines communication, is a last-mile technology that allows data to be transmitted over utility power lines. BPL is also sometimes called Internet over power line (IPL), power line communication

(PLC), or power line telecommunication (PLT). The technology uses the electrical power grid as the transmission medium, as well as the most advanced communications technology to provide high-speed broadband services over the low- and medium-voltage distribution grid. A key issue of this technology is that no new wires need to be installed in the last mile. Additionally, using the existing power grid as a transmission medium makes it possible to send and receive data through standard power sockets, and it could also be a cost-effective solution compared to other communications systems because it uses the existing power grid infrastructure.

The principle behind BPL is simple. Electricity courses over just the low-frequency portions of power lines; hence, there is 'free space' for data to stream over higher frequencies. Therefore, by installing more sophisticated computer chips into the network, they can send and receive fast data streams for more high-bandwidth applications, such as real time, and even offer new customer services, such as voice-over Internet or even video on demand.

BPL can be broadly categorized into two types—access BPL and in-house BPL [12]. The access BPL network belongs to broadband service providers. In this network, the service provider, with the help of some injection devices, injects data signals into the medium- and low-voltage power distribution network in order to provide Internet access. BPL signals may be injected onto power lines in several ways on or between different conductors. Since BPL signals cannot usually pass through an electric distribution transformer, additional equipment is usually required to allow the data signal to bypass distribution transformers, or to regenerate data, in order to get the data signal into a consumer's home. In-house BPL systems use the electrical outlets available within a building to transfer information between computers and other home electronic devices and appliances which eliminate the need for the installation of additional wires among devices, as in Fig. 1.13.

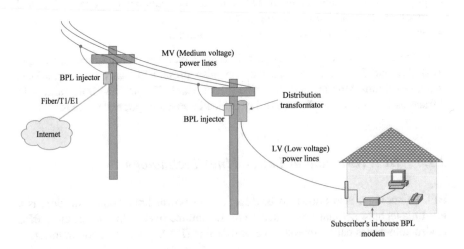

Fig. 1.13 BPL technology implementation

Initially, it was hoped that BPL technology would allow electric companies to provide high-speed access to the Internet across the last mile. In that case, the service provider would deliver phone, television, and Internet services over fiber- or copper-based long haul networks all the way to the neighborhood or curb, and then, power lines would bring the signals into the subscriber's home. The BPL subscriber would install a modem that plugs into an ordinary wall outlet and pay a subscription fee similar to that paid for other types of Internet service. No phone, cable service, or satellite connection would be required. For the reasons stated above, a lot of research and attention was given to BPL technology back in 2006, especially in the USA. The most optimistic analyses predict that the implementation of technology will have a major growth and even solve the last-mile problem, as in Fig. 1.14. Currently, depending on the architecture used, BPL can deliver access speeds between 500 kbps and 5 Mbps.

However, the actual implementation of BPL technology is bonded with a number of problems. In the first place, technology implementation requires significant upgrades of utility substations and power lines. Moreover, although it has existed for many years, this technology has not been implemented on a broad scale because of the technical difficulties involving interference. Namely, power lines are not designed to prevent radiation of RF energy, and consequently, BPL represents a significant potential interference source for all radio services using this frequency range, including the Amateur Radio Service. Electrical power lines and residential wiring in case of the BPL technology implementation act as antennas that unintentionally radiate the broadband signals as radio signals throughout entire neighborhoods and along roadsides. In practice, interference has been observed nearly one mile from the nearest BPL source. Beside this, there is also a problem related to businesses. Namely, for businesses implement this technology, Internet providers and power companies have to join in the partnership and clearly define their roles.

Proponents of the technology speculate that even if BPL has not been accepted as a viable way to deliver high-speed Internet access, it may still be used to help customers to manage their energy consumption. High-speed data transmission

Fig. 1.14 Forecast of US BPL residential subscribers (http://www.parksassociates.com)

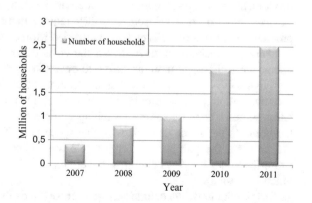

between electrical plugs in a building would allow devices such as thermostats, appliances, and smart meters to communicate with each other. Furthermore, BPL remains as an option for the implementation in rural or rugged areas.

1.1.4   FTTH Technology

The advantages of fiber as a transmission medium and the idea of introducing the fiber-optic technology in the local loop were envisaged nearly 20 years ago. Moreover, back in the 1970s, telephone and cable TV companies realized the advantages of replacing metallic cables with fiber, but the initial progress in the development of optical networks was slowed and almost halted by economical and technological factors. Due to the underdevelopment of fiber-optic technology, the cost of building a fiber-optic network was too high. The Internet was not as widespread then as it is now and customers were not ready to pay for the broadband access. As technology matured, the cost of fiber network infrastructure began to decrease, but rapid progress in development of different types of DSL technology and its widespread adoption in the nineties was another factor impeding the deployment of access network based on fiber technology. As the average transfer rate of 6 Mbps could be achieved over DSL, the cost of replacing the existing copper infrastructure with optical cables was not justified from the economic point of view.

However, the development of various applications and the rapid increase in the number of Internet users opened the door for fiber-optic technology in terms of the so-called distribution network. As part of this solution, fiber is used on trunk lines and terminated near the home, but conventional technologies are used to connect customers to the network via copper cables (twisted pair or coaxial cables). The previously described HFC networks are the example of such systems. In recent years, the Internet has become globally popular and the number of customers requiring broadband access and willing to pay for it is constantly rising. New services, such as high-definition TV, video on demand, and many other high-speed services, have been developed, and they require more bandwidth which can be provided by means of DSL or HFC networks. Moreover, the recent developments in fiber-optic technology have driven the costs down dramatically and made the fiber-optic technology more attractive to operators.

'Fiber to the home' is defined as a telecommunications architecture in which a communication path is provided over optical fiber cables extending from the telecommunications operator's switching equipment to (at least) the boundary of the home living space or business office space [13]. This communication path is provided for the purpose of carrying telecommunications traffic to one or more subscribers and for one or more services (e.g., Internet access, telephony, and/or video-television).

Despite the previous definition which understands the FTTH architecture as the 100% deployment of optical fiber in the access network, many access technologies are commonly referred to as FTTx when in fact they are simply combinations of optical fiber and twisted pair or coaxial cable networks. These technologies do not provide the inherent capability of a FTTH network. FTTx is used as a generic term for any network architecture that uses optical fiber to replace all or part of the usual copper local loop used for telecommunications. Furthermore, FTTx can be defined as a broadband telecommunications system, based on fiber-optic cables and associated optoelectronics, for the delivery of multiple advanced services, such as voice, data, and video across one link (triple-play) all the way to the home or business premises.

In today's market, there are essentially three FTTx technological options (Fig. 1.15):

- Fiber to the node/neighborhood (FTTN)—FTTN is based on fiber-optic cables run to a node serving a neighborhood. Fiber is terminated in a street cabinet up to several kilometers away from the customer's premises, with the final connection being copper. Namely, customers connect to this node through use of traditional coaxial cables or twisted pair wires;

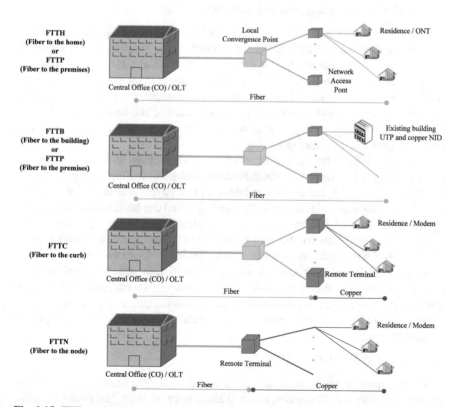

Fig. 1.15 FTTx architectures

- Fiber to the curb (FTTC)—This technology is very similar to FTTN, but the street cabinet is closer to the user's premises, typically within 300 m. Each customer has a connection to the platform via coaxial cable or twisted pair wires. Fiber to the node is often seen as an interim step toward full FTTH and is currently used by telecoms service providers to deliver advanced triple-play services;
- Fiber to the premises (FTTP)—This term is used in several contexts: as a blanket term for both FTTH and FTTB, or where the fiber network includes both homes and small businesses. In this technology, optical fiber goes directly to the end-user premises. This contrasts with FTTN, FTTC, or hybrid fiber–coaxial (HFC), which depend upon more traditional methods such as copper wires or coaxial cable for 'last-mile' delivery. FTTP can be further categorized according to where the optical fiber ends—fiber to the home (FTTH) and fiber to the building (FTTB). FTTB construction is a transitional form commonly used as a means to deliver services to existing buildings in conjunction with associated FTTH construction (e.g., for new buildings). By introducing fiber cables from the fiber termination point to the home living space or business office space, FTTB can be converted to full FTTH. Such a conversion is desirable as FTTH provides better capacity and longevity than FTTB.

The communication in FTTx architectures is defined in the following way. The optical line terminal (OLT) is the main element of the network, and it is usually placed in the central office. Optical network units (ONUs) serve as an interface to the network and are deployed on a customer's side. ONUs are connected to the OLT by means of optical fiber, and no active elements are present in the link. A single ONU can serve as point of access for one (FTTH) or multiple (FTTB or FTTC) customers and be deployed either at customer's premises (FTTH or FTTB) or on the street in a cabinet (FTTC).

FTTH is a fully optical network from the service provider to the consumer; i.e., in FTTH copper is completely absent in the outside plant and typically provides for 30–100 Mbps service, but due to the inherent characteristics of optical fiber, it can provide literally infinite bandwidth. The optical signal can be single or multiplexed signal. If multiplexed, it is brought to a splitter placed close to a group of customers. There are optical splitters of different ratios, but the most typical ratio used is 1–16. This means the multiplexed signal is split to 16 different households.

FTTH and FTTB architectures are categorized into active optical network (AON) and passive optical network (PON), involving several technical variants within each type. There are advantages and disadvantages to the deployment of AON and PON networks based on financial, bandwidth, and component considerations.

(a) Active optical network (AON)

In this type of network, there are active electrical devices (routers and switches) between the user and the central office. This is referred to as P2P (point-to-point) network. This is due to the fact that each end-user gets a dedicated fiber (or several dedicated fibers) extending from the central office. Point-to-point networks are the simplest FTTH networks to design and are characterized by the use of one fiber and laser per user. P2P networks are sometimes also referred to as All Optical Ethernet Networks (AOENs). Several examples of how P2P architectures might be deployed are shown in Fig. 1.16. The remote device or switch in the field is an active device and must be used throughout the network. The main characteristics of the P2P network include the following:

- ease to scale distance;
- active electronics in the field;
- capable to provide privacy and large-scale bandwidth portions; and
- No sharing of fiber or bandwidth for the subscriber.

On the other hand, this solution demands the installation of a lot of fibers and transmitters at CO, as well as space, and there are also density issues. Moreover, the fiber cut is difficult to repair; i.e., mean time to repair (MMTR) is low.

(b) Passive optical network (PON)

In this type of network, there are no active electrical devices between the central office and the end-user. PON is a point-to-multipoint fiber to the premises network architecture in which unpowered optical splitters are used to enable a single optical fiber to serve multiple premises, typically 16–64. A PON configuration reduces the amount of fiber and central office equipment required compared with P2P

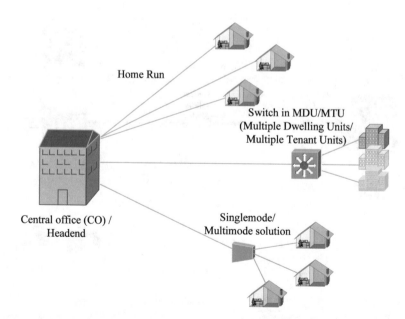

Fig. 1.16 Deployment of the P2P architectures

architectures. Although PONs can exist in three basic configurations (tree, bus, and ring), the tree topology is favored due to smaller variation in the signal power from different end station.

A PON consists of an optical line terminal (OLT) at the service provider's central office and a number of optical network units (ONUs) near end-users, as in Fig. 1.17. All transmissions in a PON are performed between OLT and ONU. The OLT resides in the central office (CO), connecting the optical access network to the metro or backbone network. The ONUs are located at the end-user location in FTTB and FTTH or in the curb in the FTTC solution.

Traffic from an OLT to an ONU is called downstream (point-to-multipoint), and traffic from an ONU to the OLT is called upstream (multipoint-to-point). In the downstream direction (from the OLT to ONUs), a PON is a point-to-multipoint network. Downstream signal coming from the central office is broadcast to each customer premises sharing a fiber. In the upstream direction, a PON is a multipoint-to-point network where multiple ONUs transmit data to the OLT. The directional properties of a passive splitter/combiner are such that one ONU transmission cannot be detected by other ONUs. However, data streams from different ONUs transmitted simultaneously still may collide. Accordingly, in the upstream direction in order to avoid collisions and fairly share the trunk fiber channel capacity and resources, a channel separation mechanism must be deployed. This mechanisms is referred to as medium access control (MAC) mechanism, and

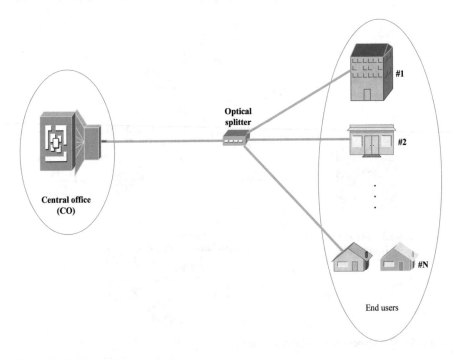

Fig. 1.17 PON architecture

currently, two approaches prevail in PON design, time-division multiple access (TDMA) and wavelength-division multiple access (WDMA) mechanisms.

Characteristics of PONs and their implementation will be further discussed in the following chapters as the most attractive solution for the realization of broadband access network.

1.1.5   Wireless Broadband

For many years now, network technologies have traditionally been based on wireline solutions that are able to support high-bandwidth data transfer and therefore numerous applications. However, these technologies do not support mobility services which have gradually opened the door for wireless technologies that are able to offer variety of services for both fixed and mobile users. The definition of wireless communications today includes communications based on satellite microwave links, terrestrial radio links, free space optical communications and even mobile cellular communications. However, in literature in general, the term 'wireless' applies only to a few microwave technologies designed for indoor or outdoor access to data networks.

Wireless broadband technologies provide ubiquitous broadband access to wireless users, enabling services that were available only to wireline users. The most common example is wireless LAN, i.e., WiFi standard, but efforts are intensively continuing to deliver broadband network access by deploying adequate radio technologies to form wireless metropolitan area networks (WMAN) and wireless wide area networks (WWAN). Broadband wireless access also represents an attractive option to operators in areas which do not have a wired access network available.

In general, wireless broadband Internet access services can be provided using mobile or fixed technologies. Wireless broadband Internet access services offered over fixed networks allow consumers to access the Internet from a fixed point and often require a direct line of sight (LOS) between the wireless transmitter and the receiver. These services have been offered using both licensed and unlicensed frequency spectrum. Today, there is a variety of technologies and standards, but this kind of wireless communication is currently dominated by the technologies defined by IEEE, where the most important ones include the IEEE 802.11 (WiFi) and IEEE 802.16 (WiMAX) family of standards [14–15], Fig. 1.18.

Nowadays, WiFi standard and its variants are well known and recognized by the telecommunications market as well as service providers and operators. WiFi (Wireless Fidelity) is a wireless local area network (WLAN) that uses specifications defined in the 802.11 standards family. The term WiFi was created by an organization called the WiFi Alliance, which oversees tests that certify product interoperability. Today, WiFi has gained acceptance in many businesses, agencies, schools, and homes as an alternative to a wired LAN. Furthermore, many airports, hotels, and other public facilities offer public access to Internet using WiFi networks.

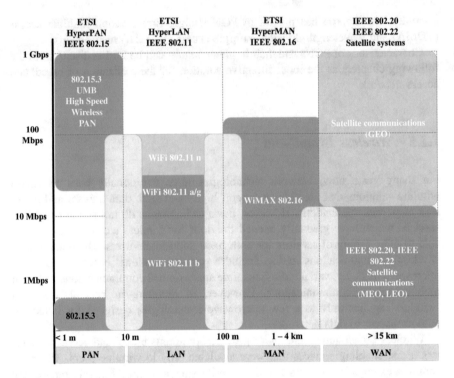

Fig. 1.18 Fixed wireless technologies

Apart from the well know group of IEEE 802.11 a/b/g standards, in today's industry a lot of attention is given to the development and implementation of the IEEE 802.11n standard. The IEEE 802.11n standard is an amendment to the IEEE 802.11-2007 wireless networking standard to improve network throughput over the two previous standards, IEEE 802.11a and IEEE 802.11g, with a significant increase in the maximum net data rate (from 54 to 600 Mbps), a slightly higher gross bit rate, slightly lower maximum throughput and the enhanced security features.

Compared with the well-known WiFi set of standards, WiMAX is a relatively new technology that is still in the process of standardization and development. WiMAX (Worldwide Interoperability for Microwave Access) is an acronym developed by the industry group called the WiMAX forum to promote the wireless data standards developed by the IEEE 802.16 working groups. In its fixed variant, 802.16d applications include last-mile wireless broadband access as an alternative to cable and DSL, wireless backhaul for metropolitan 802.11-based WLANs, and other point-to-point applications. The mobile version, 802.16e, supports roaming between base stations and therefore extends broadband service to mobile clients. Mobile WiMAX offers large coverage cell size (1–2 km in an urban environment), high download speed (around 1 Mbps at the cell edge; up to 75 Mbps under ideal conditions) and predictable QoS.

Compared with the WiFi, WiMAX provides broadband connections in greater areas, measured in square kilometers, even with links not in line of sight (NLOS), Table 1.3. Because of this, WiMAX is considered as MAN technology where 'metropolitan' refers to the extension of the areas and not to the density of population. However, comparisons (and confusion) between WiMAX and WiFi are frequent because both are related to wireless connectivity and Internet access:

- WiMAX is a long range system, covering many kilometers, which uses licensed or unlicensed spectrum to deliver connection to a network, in most cases the Internet;
- WiFi uses unlicensed spectrum to provide access to a local network;
- WiFi is more popular in end-user devices;
- WiFi runs on the CSMA/CA (Carrier Sense Multiple Access with Collision Avoidance) MAC protocol, which is connectionless and contention based, whereas WiMAX runs a connection-oriented MAC;
- WiMAX and WiFi have different QoS mechanisms:

  - WiMAX uses a QoS mechanism based on connections between the base station and the user device. Each connection is based on specific scheduling algorithms;
  - WiFi uses contention access—all subscriber stations that wish to pass data through a wireless access point (AP) are competing for the AP's attention on a random interrupt basis;

- Both IEEE 802.11 and IEEE 802.16 define P2P (Peer-to-Peer) and ad hoc networks, where an end-user communicates to users or servers on another LAN (Local Area Network) using its access point or base station. However, IEEE 802.11 also supports direct ad hoc or P2P networking between end-user devices without an access point while 802.16 end-user devices must be in range of the base station.

Table 1.3 Characteristics of wireless technologies

Standard name	Access type	Data rate— aggregate per cell	Cell radius	Frequency band
IEEE 802.11g/WiFi	WLAN	54 Mbps	50–60 m	2.4 GHz
IEEE 802.11n/WiFi	WLAN	540 Mbps	50–60 m	2.4 GHz
IEEE 802.16d (802.16-2004 Fixed WiMAX, LOS)	WMAN	32–134 Mbps (LOS)	50 km	2–66 GHz
IEEE 802.16d (802.16-2004 Fixed WiMAX, NLOS)	WMAN	75 Mbps (NLOS)	50 km	2–11 GHz
IEEE 802.16e/WiMAX	WMAN	15 Mbps for NLOS	5 km	2–11 GHz

Judging previous comparisons, it is obvious that WiFi and WiMAX are not competing technologies. While WiMAX can provide high capacity Internet access to residences and business seats, WiFi allows the extension of such connections inside the corporate site buildings and is categorized as LAN technology, Fig. 1.19. Therefore, WiMAX and WiFi are complementary technologies, defined to work together and able to ensure the best connection in accordance with user needs.

Today, WiMAX represents a powerful broadband alternative, does not require LOS, is based on a set of internationally accepted standards and spectrum set-asides, supports mobility, operates in both point-to-point mode and point-to-multipoint mode, and has the broad support of major corporations within the manufacturing and the sector of service providers. However, this technology is immature and yet to be implemented on a wider scale, especially in urban areas where its implementation is bonded with various problems [16]. Currently, WiMAX is to be considered as an alternative to or a substitute for wired connections, as backhaul for radio base stations able to bypass the PSTN (Public Switched Telephone Network) or for the Internet WiFi hot spot, and as an extension of broadband services, which are already available in urban areas, to rural areas.

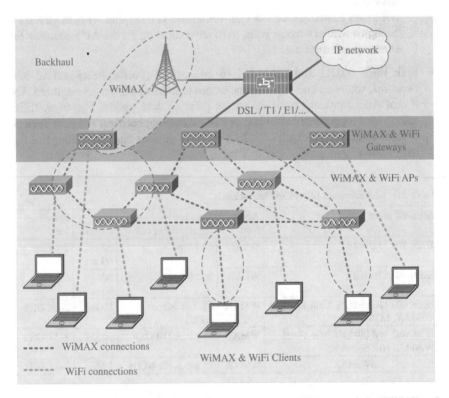

Fig. 1.19 Example of wireless network. WiMAX connections WiFi connections. WiMAX and WiFi Clients

Apart from the described wireless technologies, today's cellular communication technologies are also implemented as access technologies. Although these technologies were designed as an extension of the fixed telephone network, the constant development within this field in the past few years has made cellular communications data friendly and suitable for multimedia service delivery. As a result of constant demands for higher bandwidth, optimized coverage, and optimal spectral efficiency, most radio interface standards are moving away from CDMA (code-division multiple access)-based carriers to OFDM (orthogonal frequency-division multiplexing)-based schemes [17]. OFDM is generally more spectrally efficient than CDMA, less susceptible to interference, and offers very efficient granular bandwidth to terminals and advanced scheduling algorithms. In addition to these technologies, multiple-input multiple-output (MIMO) technology has gained a lot of attention in wireless communication since it offers significant increases in data throughput and link range without additional bandwidth or increased transmit power. It achieves this by spreading the same total transmit power over the antennas to achieve an array gain that improves the spectral efficiency (more bits per second per hertz of bandwidth) or to achieve a diversity gain that improves link reliability (reduced fading). Because of these properties, MIMO is an important part of modern wireless communication standards such as IEEE 802.11n (WiFi), 4G, 3GPP LTE (3rd Generation Partnership Project Long Term Evolution) or just LTE, WiMAX, and HSPA+ (High-Speed Packet Access).

Today, there is a number of different digital 3G and 4G cellular technologies, including WCDMA (Wideband CDMA) with HSDPA (High-Speed Downlink Packet Access), UMB (Ultra-Mobile Broadband), LTE (Long Term Evolution) Advanced, and WiMAX-Advanced (IEEE 802.16 m), as in Fig. 1.20.

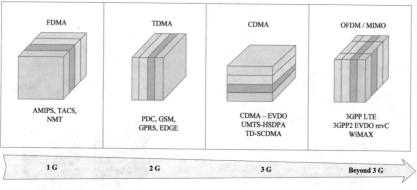

FDMA	TDMA	CDMA	OFDM / MIMO
AMIPS, TACS, NMT	PDC, GSM, GPRS, EDGE	CDMA – EVDO UMTS-HSDPA TD-SCDMA	3GPP LTE 3GPP2 EVDO revC WiMAX

| 1 G | 2 G | 3 G | Beyond 3 G |

AMPS (Advanced Mobile Phone System)
TACS (Total Access Communication System)
NMT (Nordic Mobile Telephone)

3GPP LTE (3rd Generation Partnership Project LTE)
3GPP" EVDO (3rd Generation Partnership Project 2 EVDO)
WiMAX

PDC (Personal Digital Cellular)
GSM (Global System for Mobile Communication)
GPRS (General Packet Radio Service)
EDGE (Enhaced Data Rate for Global Evolution)

CDMA-EVDO (CDMA Evolution-Data Optimized)
UMTS-HSDPA (Universal Mobile Terrestrial System With HSDPA)
TD-SCDMA (Time Division Synchronous Code Division Multiple Access)

Fig. 1.20 Multiplexing evolution of radio interface

Providers using licensed spectrum currently offer services that enable their subscribers to access the Internet with portable, 'plug-and-play' modem devices where a direct LOS between the transmitter and the receiver is not required. Customers can transport these modem devices to other locations in the provider's coverage area where a network signal is available, though they may not have the ability to maintain a connection while traveling at high speeds with handoff. Most devices are currently manufactured in accordance with vendor-specific, proprietary standards. Typical downstream speeds for portable wireless broadband services range from 768 kbps to 1.5 Mbps, and networks can extend 5–30 miles [1].

Apart from the previously described technologies, in areas where these as well as fixed broadband technologies cannot be used, satellite Internet offers Internet connection. Satellite Internet provides a connection in which the upstream data and the downstream data are sent from and arrive to the end-user through a satellite. Each subscriber's hardware includes a satellite dish antenna and a transceiver (transmitter/receiver) that operates in the microwave portion of the radio spectrum, as in Fig. 1.21.

Within the current telecommunications market, three types of satellite technologies are present: geostationary orbits (GEO), medium earth orbit (MEO), and low earth orbit (LEO). Their characteristics are summarized in Table 1.4. In a two-way satellite Internet connection, the upstream data is usually sent at speed which is slower than the speed of downstream data arrival. Thus, the connection is

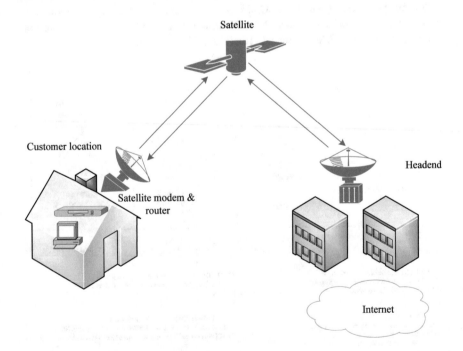

Fig. 1.21 Satellite Internet connection

Table 1.4 Characteristics of satellite technologies

Standard name	Access type	Data rate—aggregate per cell	Cell radius	Frequency band
Satellite GEO	WWAN	Up to a few Gbps	Four satellites give global coverage	3.7–8 GHz (C Band), 10–18 GHz (Ku Band), 18–31 GHz(Ka Band), 37.5–50.2 GHz (Q/V Band)
Satellite MEO	WWAN	Up to a few Mbps	11 satellites give global coverage	Same as GEO satellites
Satellite LEO	WWAN	Up to a few Mbps	Varies	Same as GEO satellites

asymmetric. Uplink speeds are nominally 50–150 kbps for a subscriber using a single computer. The downlink occurs at speeds ranging from about 150 kbps to more than 1200 kbps, depending on factors such as Internet traffic, the capacity of the server, and the sizes of downloaded files. The offered rates are not sufficient for the transmission of multimedia applications, and the installation is expensive in comparison with other technologies.

Geostationary satellites can offer higher data speeds, but their signals cannot reach some polar regions of the world. Different types of satellite systems have a wide range of different features and technical limitations which can greatly affect their usefulness and performance in specific applications [1]. Compared with ground-based communications, all geostationary satellite communications experience high latency due to the fact that signal has to travel 35,786 km to the satellite in geostationary orbit and back to Earth again. Even at the speed of light (about 300,000 km/s or 186,000 miles per second), this delay can be significant. Even if all other signaling delays were eliminated, it would still take a radio signal about 250 ms, or about a quarter of a second, to travel to the satellite and back to the ground. For an Internet packet, that delay is doubled before the reply is received and this is the theoretical minimum. If we add other normal delays from different network sources, it gives a typical one-way connection latency of 500–700 ms from the user to the ISP or about 1000–1400 ms latency for the total round-trip time (RTT) back to the user. This is substantially more than what most dial-up users' experience, a typical 150- to 200-ms total latency.

Internet latency mentioned above makes satellite Internet service problematic for applications requiring real-time user input, such as online games. This delay can also be irritating with interactive applications, such as VoIP (voice-over IP), videoconferencing, or other person-to-person communications. It will cause the common market applications (such as Skype) to behave unpredictably and fail, as these are not designed for the difficult compensation required for the high-latency connections.

For geostationary satellites, there is no way to eliminate latency, but the problem can be somewhat mitigated in Internet communications with TCP (Transmission

Control Protocol) acceleration features that shorten RTT per packet by splitting the feedback loop between the sender and the receiver. Such acceleration features are usually present in the recent technology developments embedded in new satellite Internet services. MEO and LEO satellites do not have such great delays. The current LEO constellations have delays of less than 40-ms RTT, but their throughput is less than broadband at 64 kbps per channel. The different variants of MEO constellation are currently under development, and the future design should bring much higher throughput with links well in excess of 1–1.2 Gbps with a reduced latency of approximately 7 ms. However, satellite Internet systems are sometimes the only option for people in rural areas where other broadband (cheaper) connections are not available. A satellite installation can be used even where the most basic electricity grid utilities are lacking, given that there exists a generator or battery power supply that can produce enough electricity to run a desktop computer system.

1.2   Next-Generation Networks and Service Evolution

Over the past few years, Internet service providers have begun to use new strategies to realize next-generation network (NGN) called triple-play networks because they allow simultaneous transmission of voice, video, and data signals [1]. Triple-play services usually mean voice, data, and video bundled over a broadband connection. Moreover, NGN can be thought of as a packet-based network where the packet switching and transport elements are logically and physically separated from the service control intelligence. The control intelligence supports transmission of all types of services over the packet-based transport network, including basic voice telephony services, data, video, multimedia, advanced broadband, and management applications, which can be thought of as just another type of service, as in Fig. 1.22. These networks represent the ongoing goal of network operators to gain maximum revenue leverage for their investment in infrastructure, and therefore, protecting that investment is critical. NGN network should provide multiple services and multiple devices but at the same time—one network, one vendor, and one bill.

Basic characteristics of triple-play networks include the following [1]:

- Various services: Internet access; telephone/video calls; video on demand; mobile connectivity.
- Converged network: IP infrastructure; packet-oriented networks; implemented a quality of service; multiservice environment.
- Using the services of one service provider: One bill for all services; unique customer support for all services; integrated voice mail, and address book.

Internet traffic is expected to continue to grow rapidly in the years to come, as in Fig. 1.23. To be able to cope with this, service providers must have a network that is simple, economical, and scalable. In addition, there are rising needs for services

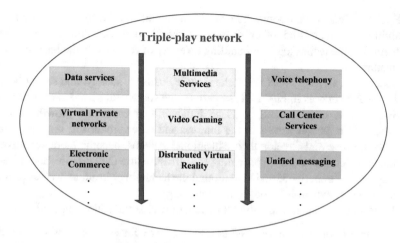

Fig. 1.22 Services in a triple-play network

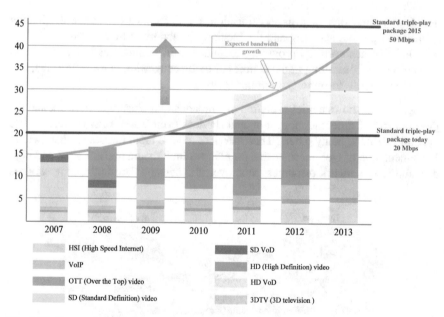

Fig. 1.23 Expected worldwide (continuous) bandwidth growth. *Source* FTTH council; operators: Alcatel-Lucent

that emphasize quality, especially video quality and security. Moreover, new services that cross conventional service boundaries constantly emerge. The convergence of services and the emergence of new business models require a network that can flexibly adapt to a new range of applications.

From a technical standpoint, there has been dramatic progress in the capacity and reliability of routers and other devices in the IP (Internet Protocol) networks. Furthermore, the technology for handling a variety of services on the basis of the IP has matured.

Application areas of NGN networks include business, society, and subscribers in the access network, as in Fig. 1.24. Moreover, the successful adoption of broadband services is also often looked at as an indication of economic growth in some countries. With greater access to the Internet and other broadband services, countries can improve and develop their educational systems, become more attractive to multinational corporations by providing the infrastructure they require, and develop a strong pool of experience and skill around the technology itself, as well as provide the information available through the technology.

Under such circumstances, the NGN must provide the following [18]:

- open interfaces that can be used to interconnect with various providers and can provide the basis for the joint creation of services;
- QoS support for applications that require high communication quality, such as video and voice applications;
- security functions such as identification and authentication of the originating party to prevent spoofing and the blocking of abnormal traffic; and
- reliability achieved by providing redundancy in communication circuits and equipment, controlling traffic at times of traffic congestion in a specific area, and giving priority to important communication.

Figure 1.25 shows a brief evolution of a broadband multiservice network and the development of different application and services from VoIP, VPN (virtual private vetwork), VoD to many others. These services require rapid increase in bandwidth, implementation of QoS mechanisms for efficient multimedia transmission, and high availability of network resources. Moreover, the demand for multicasting services

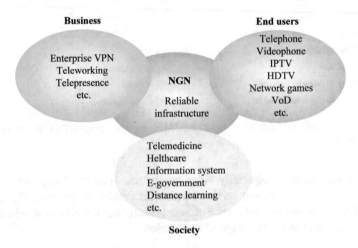

Fig. 1.24 Development of broadband services

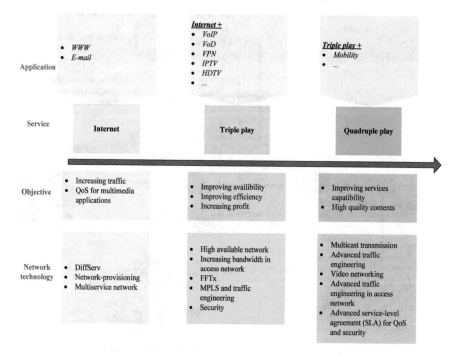

Fig. 1.25 Next-generation networks development

with high-quality and specialized contents is also expected to increase. Hence, new services will require a reliable, secure, and highly available network. In order to provide this service, providers have implemented WDM technology in the back-bone along with MPLS (Multiprotocol Label Switching) and advanced traffic engineering techniques.

On the other hand, in order to maximize service revenues and minimize sub-scriber turnover, service providers have to offer a complete set of bundled triple-play services to residential subscribers that include voice, high-speed Internet, broadcast TV, and VoD.

The 'new broadband' user requests from a service provider a much wider range of services than it has until recently been the case, as in Fig. 1.26. In addition to basic data transfer, service providers in the access network must now provide the band-width that will be able to support the transmission of different applications such as VoD and HDTV, while at the same time, it must be able to provide the appropriate service quality to end-users. Moreover, a variety of services (some already available, others still at the conceptual stage) have been linked to NGN initiatives and con-sidered to be the likely candidates for NGN implementations. While some of these services can be offered on existing platforms, others benefit from the advanced control, management, and signaling capabilities of NGNs. Although the emerging and new services are likely to be the strongest drivers for NGNs, most of the initial NGNs profits may actually result from the bundling of traditional services.

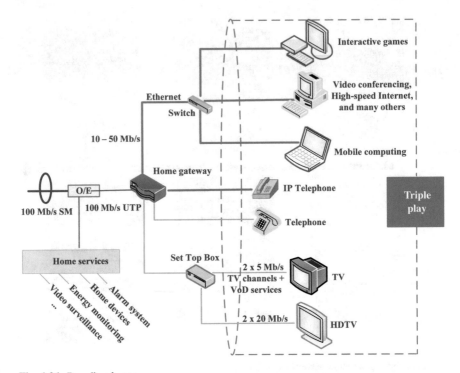

Fig. 1.26 Broadband user

Future NGNs will most likely enable a much broader array of service types, including the following:

- specialized resource services (provision and management, multimedia multi-point conferencing bridges, media conversion units, voice recognition units, etc.);
- processing and storage services (provision and management of information storage units for messaging, file servers, terminal servers, OS platforms, etc.);
- middleware services (naming, brokering, security, licensing, transactions, etc.);
- application-specific services (business applications, e-Commerce applications, supply-chain management applications, interactive video games, etc.);
- content provision services that provide broker information content (e.g., electronic training, information push services, etc.);
- interworking services for interactions with other types of applications, services, networks, protocols, or formats (e.g., EDI translation); and
- management services to maintain, operate, and manage communications/computing networks and services.

Currently, the most prevalent technologies in the access network are based on DSL and cable networks, as in Fig. 1.27. Even though we witnessed a rapid development of these technologies as well as wireless technologies in the last decade, they cannot provide enough bandwidth for delivering triple-play services to end-users [19].

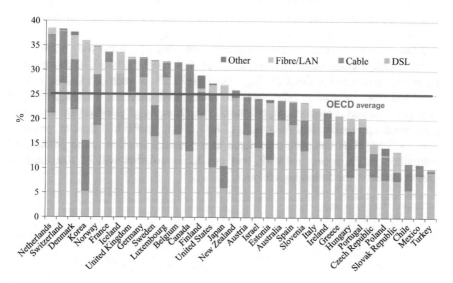

Fig. 1.27 Fixed (wired) broadband subscriptions per 100 inhabitants, by technology, June 2011. *Source* OECD

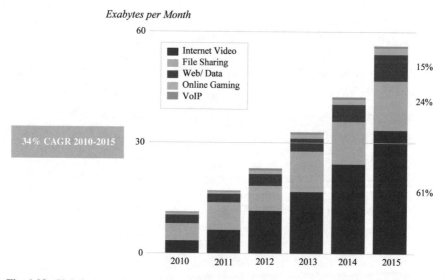

Fig. 1.28 Global consumer Internet traffic (http://www.cisco.com/en/US/solutions/collateral/ ns341/ns525/ns537/ns705/ns827/white paper c11-481360_ns827_Networking_Solutions_White_ Paper.html)

To resolve the bandwidth bottleneck issue, service providers bring optical fiber in the first mile in the DSL as well as cable networks. In the DSL-based networks, remote DSLAMs use fiber-optic links to connect with the central office. In cable networks, optical nodes are deployed close to the end-users.

However, it is expected that the future will bring continued convergence across residential, business, and mobile services. The evolution of services in the access network brings new expansion called a quadruple-play network that offers mobility in addition to voice, data, and video already provided by a triple-play network.

In order to accommodate new services, it is vital that the network be able to scale up to tens and even hundreds of Gbps, as in Fig. 1.28. In such conditions, service providers must be able to offer customers integrated multimedia applications for both business and entertainment, but the existing broadband solutions are unable to provide enough bandwidth. As a result of new developments and the emergence of new services, PON has been considered as a potential solution to the problem of congestion in the access network, and this likely solution will be discussed in greater detail in the following chapters.

Chapter 2
Quality of Service Implementation

2.1 QoS Definition

In order to transfer the various applications and services that exist in the triple-play network, it is necessary to provide resources and different levels of performance so that the end-users would get a satisfactory level of service. Within a converged network, QoS is by far the most important implementation consideration. QoS is a networking term that specifies a guaranteed network data performance level. QoS guarantees are important if the network capacity is insufficient, especially for real-time streaming multimedia applications such as VoIP, online games, and IP-TV, since these often require a fixed bit rate and are delay sensitive. Accordingly, QoS could also be defined as a set of algorithms and forms which enable different priorities to be given to different applications, users, or data flows, or to guarantee a certain level of performance for data flow. In practical terms, QoS can be described as a mechanism that has to provide transmission of audio and video data with minimum delay and minimum packet loss rate. Without QoS configuration, transmission of IP voice or videoconferencing calls would be unreliable, inconsistent, and often unsatisfactory.

The performance of a packet-based network can be characterized by several parameters where the most important ones include the following [20]:

- Bandwidth—typically specified in kilo or mega bits per second (kbps or Mbps), is measured as the average number of bits per second that can travel successfully through the network;
- The average packet delay (latency)—End-to-end delay is the average time it takes for a network packet to traverse the network from one endpoint to the other;
- The delay variation (jitter)—Jitter is the variation in the end-to-end delay of sequential packets;
- The percentage of lost packets (packet loss ratio)—Packet loss is measured as the percent of transmitted packets that never reach the intended destination.

© Academic Mind and Springer International Publishing AG 2017
M. Radivojević and P. Matavulj, *The Emerging WDM EPON*,
DOI 10.1007/978-3-319-54224-9_2

Table 2.1 Characteristics of different types of traffic without the implemented QoS mechanisms

Type of traffic	Performance without QoS mechanisms
Speech	Speech is difficult to understand
	Interruptions in conversation
	Delays conversation; the participants do not know when the other party is closed
	Calls are disconnected
Video	The image is blurry
	Speech and images are not synchronized
	Slow moving
Data	The data arrives at their destination with delay
	Users give up using the service due to the excessive response time
	The network is slow

For example, in case of IP voice and video communications systems, the bandwidth should be as large as possible while the end-to-end delay, jitter, and packet loss are minimized in order for systems to work properly. Lower end-to-end delay leads to a more satisfactory, natural feeling conferencing experience, while large delay values lead to unnatural conversations with long pauses between phrases or sentences. Moreover, large jitter values may cause packets to arrive in the wrong sequence causing jerky video or stuttering audio.

Table 2.1 shows the characteristics of different types of traffic without the implementation of QoS mechanisms. It is obvious that QoS support has become the essential parameter for the successful realization of multiservice networks.

2.2 QoS Parameters

QoS implementation should increase the performance of traffic that is transmitted in the network, but the fact that improving the characteristics of one type of traffic can cause the loss of performance transmission of another type of traffic must be taken into account as well. Successful implementation of QoS includes the basic setup of the parameters given above (bandwidth, average packet delay, jitter, and packet loss) in order to improve network performance and meet the demands of end-users. The parameters are defined as follows [20]:

A. *Bandwidth*

In the IP-based networks, bandwidth is defined as the capacity of the transmission medium. It is usually referred as 'speed' or network throughput and specified in kbps, Mbps, or Gbps. In case of QoS implementation, this is related to bandwidth allocation techniques because QoS techniques cannot affect or increase the actual bandwidth of a link. Specifically, QoS mechanisms cannot create additional bandwidth, but they can contribute to a more efficient allocation of the existing bandwidth.

B. *Packet delay*

The delay of a network specifies how long it takes for a bit of data to travel across the network from one node or endpoint to another. Although users only care about the total delay of a network, designers and engineers need to conduct precise measurements. Thus, engineers usually report both the maximum and average delay, and the following components have been defined (Fig. 2.1):

- *Processing delay*—the time that elapses between the receipt of data on the input port and its forwarding to the destination on the output port. This delay depends on the CPU speed and CPU load in the system;
- *Serialization delay*—the time it takes for the packet or frame to be transmitted, i.e., time it takes to push the packet's bits onto the link. This delay depends on the speed of the communication link;
- *Queuing delay*—the time that the packet spends in routing queues, i.e., the delay between the point of entry of a packet in the transmit queue and the actual point of transmission of the message. Network interfaces transmit one frame at a time (typically one bit at a time). Hence, when two or more packets are forwarded to a network interface at the same time or close to the same time, one packet is transmitted while the others are put in a queue on the interface buffer to await their turn at the interface. Packets that are put into the queue must wait until they can be transmitted, adding milliseconds to the delay. Increases in queue delay can be measured and detected by monitoring traffic along a given network path. This delay depends on the load on the communication link;
- *Propagation delay*—This is the delay between the point of transmission of the last bit of the packet and the point of reception of the last bit of the packet at the other end. This delay depends on the physical characteristics of the communication link.

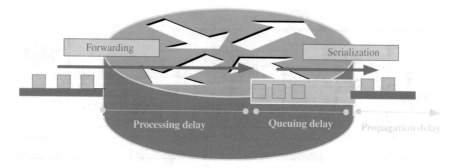

Fig. 2.1 Definition of delay in IP networks

The term end-to-end delay is used as a sum of all propagation, processing, serialization, and queuing delays in the path as shown in Fig. 2.2. Jitter defines the variation in the delay. In best-effort networks, propagation and serialization delays are fixed, while processing and queuing delays are unpredictable.

Most of the processing, queuing, and propagation delays are influenced by the following factors:

- Average length of the queue;
- Average length of packets in the queue;
- Link bandwidth.

There are several approaches to accelerate packet dispatching of delay-sensitive flows:

- Increase link capacity. Enough bandwidth causes queues to shrink, making sure that packets do not wait long before they can be transmitted. Additionally, more bandwidth reduces serialization time. On the other hand, this might be an unrealistic approach due to the costs associated with the upgrade;
- A more cost-effective approach is to enable a queuing mechanism that can give priority to delay-sensitive packets by forwarding them ahead of other packets.

Today, a variety of queuing mechanisms are used in IP networks, such as priority queuing, custom queuing, and weighted fair queuing. By minimizing delay, jitter is also reduced.

C. *Packet delay variation (jitter)*

In computer networking, packets from the source will reach the destination with different delays [21]. A packet's delay varies with its position in the queues of the routers along the path between the source and destination, and this position can vary unpredictably. This variation in delay is known as jitter and can seriously affect the

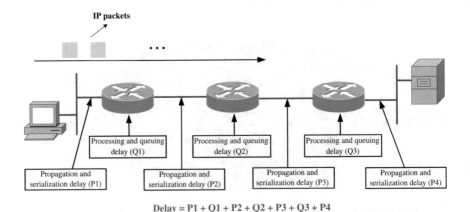

Delay = P1 + Q1 + P2 + Q2 + P3 + Q3 + P4

Fig. 2.2 Calculation of end-to-end delay

quality of streaming audio and/or video. It is important to emphasize that with delay minimization, jitter is also reduced (the delay is more predictable).

D. *Packet loss*

Packet loss in the network is defined as the percentage of lost packets on the path from source to destination due to the packet overload of the buffer, line errors, or even because of the implemented QoS mechanisms. Namely, the routers might fail to deliver some packets if their data is corrupted or if they arrive when their buffers are already full. Moreover, the receiving application may ask for this information to be retransmitted, possibly causing severe delays in the overall transmission. Packet loss is usually the result of congestion on an interface. Most applications that use TCP are slowed down due to TCP adjusting to the network's resources. There are some other applications that do not use TCP and cannot handle congestion.

The following approaches can be taken to prevent drops of sensitive applications:

- Increased link capacity to ease or prevent congestion;
- Guarantee enough bandwidth and increase buffer space to accommodate bursts of fragile applications.

There are several mechanisms that can guarantee bandwidth and/or provide prioritized forwarding of drop sensitive applications such as priority and custom queuing In order to decrease packet loss rate, it is necessary to prevent congestion (by dropping other packets before congestion occurs).

Table 2.2 summarizes the QoS requirements, i.e., expected values of the parameters of different multimedia applications.

In addition to the previous definitions and parameters, QoS implementation in the IP network also includes the following definitions:

Table 2.2 QoS requirements of different multimedia applications

Application	Bandwidth	Delay	Jitter	Packet loss rate
Voice transmission	Low-Medium	Low	Low	Low
Interactive voice transmission	Medium	Low	Low	Low
Streaming video traffic	Medium-High	High	High	Low
Video signalization	Low	Low	Medium	Medium
Speech signalization	Low	Low	Medium	Medium
Interactive and criminal record	Low-Medium	Low-Medium	Low-Medium	Medium-High
Non-interactive critical data	Variable, usually High	High	High	Medium
Interactive, non-critical	Variable, usually Medium	High	High	Medium

- Flow (or microflow) is a sequence of packets identified by source and destination IP addresses, protocol identifier (e.g., TCP and UDP), and source and
  destination port numbers;
- Traffic stream is a collection of flows with a common set of parameters (e.g., the
  same port number and the same source and destination network);
- Traffic profile specifies typical properties of a traffic stream (average rate and
  burstiness). Stream provisioning should be performed based on traffic profiles
  and the importance of traffic streams.

2.3 QoS Implementation

The key point for providing QoS in the NGN networks is the ability of these
networks to differentiate traffic and to provide differentiated service levels based on
the types of traffic. Namely, for the successful transmission of real-time applications, such as VoIP, the amount of available bandwidth and low end-to-end delay is
necessary. At the same time, applications such as file transfer and/or e-mail
transmissions are quite insensitive to bandwidth and delay issues.

The standard IP architecture is based on a 'best-effort' model where all network
traffic is equally important and everyone receives service based on availability,
without any guarantees [22]. Best-effort delivery describes a network service in
which the network does not provide any guarantees that data will be delivered or
that a user will be given a guaranteed QoS level or a certain priority. The implementation of QoS mechanisms is not supported, and users obtain best-effort service
depending on the current traffic load. However, an important issue regarding the
Internet, and consequently every network connected to it, is the support for multimedia applications. In networks with capacity problems that have to transmit
real-time multimedia applications such as IP telephony and IP-TV, the best-effort
mechanism is not able to provide enough resources. In this situation, mechanisms
that are able to provide differentiated or guaranteed QoS to certain data flows must
be employed.

The IETF (Internet Engineering Task Force) is responsible for standardization of
the Internet and most of the protocols are used in the Internet. When faced with a
challenge, vendors introduce their own solutions. However, the IETF is there to
create standards that allow different vendor's equipment to interoperate. One of the
challenges in the past was to introduce QoS into the best-effort-driven Internet.
Accordingly, two QoS architectures have emerged in the IETF:

- Integrated services architecture (IntServ), which provides end-to-end QoS on a
  per-flow basis and incorporates end-to-end signaling;
- Differentiated services architecture (DiffServ), which supports QoS for traffic
  aggregates through the implementation of different QoS for different traffic
  classes.

2.3.1 IntServ Model

IntServ model, or integrated services model defined in IETF RFC 1633 [23], represents an architecture that specifies the elements which will ensure QoS on networks and can be described as a flow-based system. In this model, every router in the system implements IntServ, and every application that requires some kind of guarantees has to make an individual reservation, Fig. 2.3. The integrated services model expects applications to signal their QoS requirements to the network. In order to accommodate such requirements, network devices will use this information to manage network resources. Network devices control network resources and provide QoS services by configuring the traffic handling mechanisms. Resource reservation can be applied to individual flows or aggregated flows.

Resource reservation mechanisms include the following functions:

- Provision of resource reservation signaling that notifies all devices along the communication path on the multimedia applications' QoS requirements;
- Delivery of QoS requirements to the admission control mechanism that decides whether there are available resources to meet the new QoS requirements;
- Notification of the application regarding the admission result.

Applications use the transport layer RSVP (Resource Reservation Protocol) to request and reserve resources through a network. RSVP does not transport application data but is similar to a control protocol. Moreover, this protocol assumes that resources are reserved for every flow requiring QoS at every router hop in the path between the receiver and transmitter using end-to-end signaling. RSVP requests resources for simplex flows (a traffic stream in only one direction from sender to

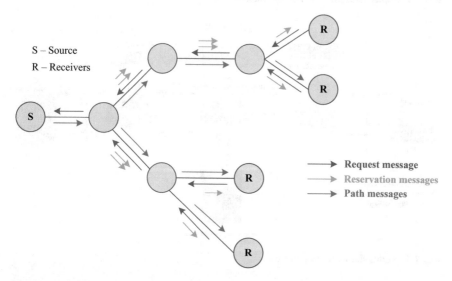

Fig. 2.3 RSVP functions

one or more receivers) and is receiver-oriented (the receiver of a data flow initiates and maintains the resource reservation for that flow).

RSVP operates over an IPv4 and provides receiver-initiated setup of resource reservations for data flows. RSVP's main functionality is to exchange QoS requirement information between the source host, the destination host, and inter-mediate devices as shown Fig. 2.4.

The IntServ model itself describes the application of QoS in IP networks. In addition to RSVP, additional standards were developed to cover the exact protocols used for QoS implementation (Fig. 2.5):

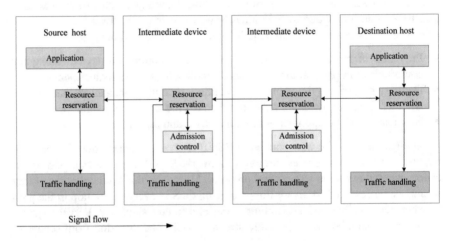

Fig. 2.4 Implementation of RSVP mechanism

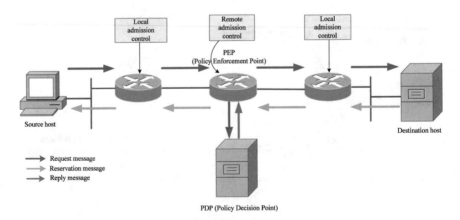

Fig. 2.5 Additional QoS mechanism

- Admission control is either implemented locally on the routers or offloaded to central servers;
- COPS (Common Open Policy Service) is another IETF standard that defines a protocol that can be used for policy exchange between network devices, PEPs (Policy Enforcement Points), and policy servers, PDPs (Policy Decision Points). An additional standard was added to integrate RSVP with COPS. The COPS (Common Open Policy Service) protocol is defined in RFC 2748 [24]. COPS usage for RSVP is defined in RFC 2749 [25].

To enhance the RSVP functionality, admission controlmechanisms have to be implemented in order to determine whether the application (flow) can obtain the requested resources. Admission control mechanisms allow a network to reject (or downgrade) new RSVP sessions if one of the interfaces in the path has reached the limit (all reserved bandwidth is booked). These mechanisms will ensure that QoS parameters of the existing sessions will not be degraded and the new session will be provided with adequate QoS support.

In case the network does not have the resources necessary for the successful transfer of new flow, the mechanism may reject a request to establish a new session or to accept it and it will send a message to the end-user that the session is accepted but that the network may not be able to provide sufficient resources. Admission control mechanisms and resource reservation signaling mechanisms closely cooperate with each other and are implemented in the same device.

In today's networks, there are two approaches for the implementation of the admission control:

- Explicit admission control.
  This approach is based on explicit resource reservation. Applications send the request that contains QoS parameters to join the network through the resource reservation signaling mechanism. The request is forwarded to the admission control mechanism that decides to accept or reject the application based on the application's QoS requirements, available resources, performance criteria, and network policy;
- Implicit admission control.
  There is no explicit resource reservation signaling. The admission control mechanism relies on bandwidth overprovisioning and traffic control (i.e., traffic policing).

In addition to the admission control mechanisms, the COPS protocol is implemented in order to enhance the functionality of RSVP. The COPS protocol has been specified for RSVP by RAP (Resource Allocation Protocol) working group of IETF. The protocol is part of the Internet Protocol suite as defined by the IETF's RFC 2748. It is a simple query and response protocol that can be used to exchange policy information between a policy server (PDP) and its clients (PEPs). One example of a policy client is an RSVP router that must exercise policy-based admission control over RSVP usage.

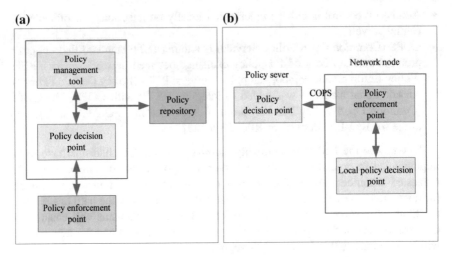

Fig. 2.6 a Intra-domain policy management; **b** Common open policy service

The COPS protocol specifies a simple client/server model for supporting policy control over QoS signaling protocols (e.g., RSVP). Policies are stored on servers, and acted upon by PDP, and are enforced on clients (PEP) as shown Fig. 2.6. At least one policy server exists in each controlled administrative domain. COPS protocol has a simple but extensible design.

The main characteristics of COPS include the following [26]:

- The protocol employs a client/server model where the PEP sends requests and updates to the remote PDP and deletes and the PDP returns decisions back to the PEP;
- The protocol uses TCP as its transport protocol for the reliable exchange of messages between policy clients and a server. Therefore, no additional mechanisms are necessary for the reliable communication between a server and its clients;
- The protocol is extensible in that it is designed to leverage off self-identifying objects and can support diverse client-specific information without requiring modifications to the COPS protocol itself. The protocol was created for the general administration, configuration, and enforcement of policies;
- COPS protocol provides message level security for authentication, replay protection, and message integrity. The protocol is stateful in two main aspects:

  – Request/Decision state is shared between client and server—Requests from the client PEP are installed or remembered by the remote PDP until they are explicitly deleted by the PEP. At the same time, decisions from the remote PDP can be generated asynchronously at any time for a currently installed request state;

- State from various events (Request/Decision pairs) may be inter-associated—
  The server may respond to new queries differently because of the previously
  installed Request/Decision state(s) that are related.

Additionally, the protocol is stateful in that it allows the server to push configuration information to the client, and then allows the server to remove such state
from the client when it is no longer applicable.

From the point of view of the client or PEP, there are two models of COPS: the
outsourcing model and the provisioning model.

- The outsourcing model is the simplest COPS implementation. In this model, all
  policies are stored at the PDP. Whenever the PEP needs to make a decision, it
  sends all relevant information to the PDP. The PDP analyzes the information,
  makes the decision, and relays it to the PEP. The PEP then simply enforces the
  decision;
- In the provisioning model, COPS is used for policy provisioning (COPS-PR),
  and the PEP reports its decision-making capabilities to the PDP. The PDP then
  downloads relevant policies to the PEP. The PEP can then make its own
  decisions based on these policies. The provisioning model uses the policy
  information base as a repository for the policies.

The main benefits of RSVP are as follows:

- It signals QoS requests per individual flow. The network can then provide
  guarantees to these individual flows. However, in the same time, the problem of
  RSVP is that it does not scale to large networks because of the large numbers of
  concurrent RSVP flows;
- It informs network devices of flow parameters (IP addresses and port numbers).
  Some applications use dynamic port numbers, which can be difficult for network
  devices to recognize. Hence, the additional mechanism must be used to sup-
  plement RSVP for applications that use dynamic port numbers but do not use
  RSVP.

The main drawbacks of RSVP are as follows:

- Continuous signaling due to the stateless operation of RSVP;
- RSVP is not scalable to large networks where per-flow guarantees would have
  to be made to thousands of flows.

The major advantage of IntServ is that it provides service classes which closely
match different application types. However, although IntServ is a straightforward
QoS model, end-to-end service guarantees cannot be supported unless all nodes
along the route support IntServ. Therefore, scalability is a key architectural concern
for IntServ, since it requires end-to-end signaling and must maintain a per-flow state
at every router on the path.

One way of solving this problem is by using a multilevel approach, where
per-microflow resource reservation (i.e., resource reservation for individual users) is
done in the edge network, while in the core network, resources are reserved for

aggregate flows only. The routers that lie between these different levels must adjust the amount of aggregate bandwidth reserved from the core network so that the reservation requests for individual flows from the edge network can be better satisfied. This approach is known as a hybrid model RSVP-DS (Resource Reservation Protocol with DiffServ) and is implemented as a backbone service architecture concept.

2.3.2 DiffServ Model

Differentiated services or DiffServ is a computer networking architecture that specifies a simple, scalable, and coarse-grained mechanism for classifying, managing network traffic, and providing QoS guarantees on modern IP networks [20].

The architecture describes services and allows for more user-defined services to be used in a DiffServ-enabled network. DiffServ operates on the principle of traffic classification, where each data packet is placed into a limited number of traffic classes, rather than differentiating network traffic based on the requirements of an individual flow. A class can be identified as a single application or it can be identified based on source or destination IP address (the most common case). Each traffic class can be managed differently, ensuring privileged treatment for higher priority traffic on the network. DiffServ recommends a standardized set of traffic classes for ensuring simpler interoperability between different networks and different vendors' equipment.

The idea behind this framework is for the network to recognize a class without having to receive any requests from applications as it is the case within the Intserv model. This allows the QoS mechanisms to be applied to other applications that do not have the RSVP functionality, which is the case with the majority of applications that use IP protocol. DiffServ can, for example, be used to provide low latency to critical network traffic such as voice or video, while providing simple best-effort traffic guarantees to non-critical services such as Web traffic or file transfers.

In DiffServ model, ToS (Type of Service) field is now called the DS (Differentiated Services) field (RFC 2474 [27]). The DS field consists of a differentiated services code point (DSCP) field (6 bits) and explicit congestion notification (ECN) field, which occupies the least significant two bits, Fig. 2.7. The six bits replace the three IP precedence bits, and is called the DSCP but maintains interoperability with non-DS compliant devices (those that still use IP precedence). Because of this backward compatibility, DiffServ can be gradually deployed in large networks.

Namely, DiffServ uses the 6-bit DSCP field in the header of IP packets for packet classification purposes as shown in Fig. 2.7. DSCP replaces the previously used IP precedence, a 3-bit field in the ToS byte of the IP header originally used for classifying and prioritizing types of traffic. In contrast to IntServ, which is defined as a flow-based mechanism, DiffServ is a class-based mechanism for traffic

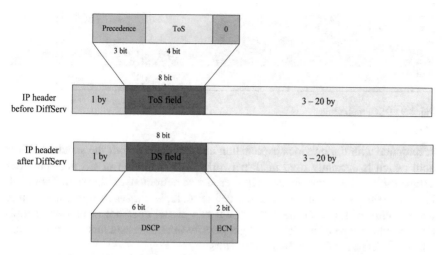

Fig. 2.7 ToS field

management which provides a framework for classification and differentiated treatment. Unlike the IP precedence solution, the ToS byte is completely redefined.

With DSCP, in any given node, up to 64 different aggregates/classes can be supported (2^6), which allows finer traffic segregation compared to IP precedence and eight supported traffic classes (2^3). Every classification and QoS revolves around the DSCP in the DiffServ model. The 64 possible values of DSCP field are divided into three groups (DSCP pools). The two groups that comprise 32 possible DSCP values are reserved for experimental use. The third group includes the other 32 values, where 21 values are defined in the DiffServ specification and are to be used when forwarding traffic as shown in Fig. 2.8.

Furthermore, the DSCP field is divided into two sub-fields of the three-bit field, called the class selector (CS) and the drop probability (DP) field that defines the probability of rejection of a given packet within a given class in the event of congestion in the network as shown in Fig. 2.9. DS0 bit of DP field is set to zero in

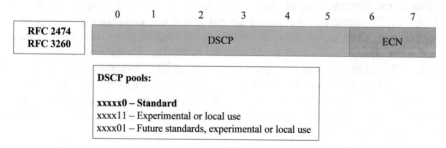

Fig. 2.8 DiffServ pool

ToS byte	P2	P1	P0	T3	T2	T1	T0	ZERO
DS byte	DS5	DS4	DS3	DS2	DS1	DS0	ECN1	ECN0
	Class selector (CS)			Drop probability (DP)				

Fig. 2.9 DSCP definition

accordance with the previous explanation and the definition of the standard DSCP pool which is currently used in IP networks. The definition of CS field provides compatibility with earlier models (IP Precedence definition) as shown in Table 2.4. However, the use of the DS field can define multiple classes of traffic and a finer classification of traffic in the network but this model does not incorporate definitions of which type of traffic should get priority treatment, and this is consequently left to be determined by network operators.

Each router on the network is configured to differentiate traffic based on its class. The collection of packets that have the same DSCP value (also called a code point) in them, and which are crossing in a particular direction, is called a behavior aggregate (BA). Packets from multiple applications/sources could belong to the same BA.

Furthermore, DiffServ-aware routers implement per-hop behaviors (PHBs), which define the packet forwarding properties associated with a certain class of traffic. Generally, traffic may be classified by many different parameters, such as source address, destination address, or traffic type and assigned to a specific traffic class. According to the standard (RFC 2474), the PHB is determined by the differentiated services (DS) field of the IPv4 header. Now, BA identifies packets marked with the same DSCP where PHB is applied to each BA according to the QoS policy.

However, the RFC standards do not dictate the manner of implementing PHBs, and it is the responsibility of the vendor. In more concrete terms, a PHB refers to the packet scheduling, queuing, policing, or shaping behavior of a node on any given packet belonging to a BA, as configured by a SLA (service-level agreement) or policy. For example, a vendor can implement queuing techniques that can base their PHB on the IP precedence or DSCP value in the IP header of a packet. Based on DSCP or IP precedence, traffic can be put into a particular service class where packets within this service class are treated in the same way. As we have explained earlier, in theory, the network could have up to 64 (i.e., 2^6) different traffic classes using different markings in the DSCP. The DiffServ RFC standards recommend, but do not require, certain encodings which give a network operator great flexibility in defining traffic classes. To date, four standard PHBs have been available for constructing a DiffServ-enabled network and achieving coarse-grained, end-to-end CoS (class of service) and QoS (Fig. 2.10):

DSCP encoding

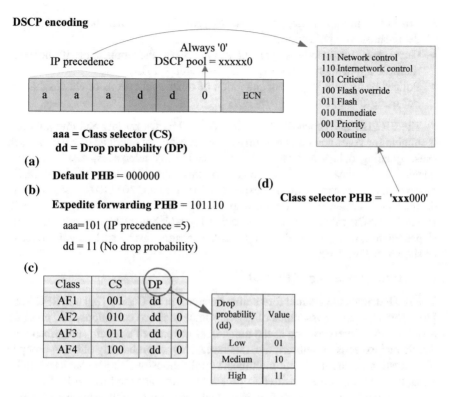

Fig. 2.10 PHB definition

- Default PHB—which is typically best-effort traffic;
- Expedited forwarding (EF) PHB—dedicated to low-loss, low-latency traffic;
- Assured forwarding (AF) PHB—gives assurance of delivery under prescribed conditions;
- Class selector PHBs—which maintain backward compatibility with the IP precedence field.

A. *Default PHB*

A default PHB is typically best-effort traffic. Any traffic that does not meet the requirements of any of the other defined classes is classified as the default PHB. In most cases, the default PHB has best-effort forwarding characteristics. The recommended DSCP for the default PHB is '000000' (in binary) as shown in Fig. 2.10a.

B. *Expedited forwarding (EF) PHB*

EF PHB is dedicated to low latency, low packet loss traffic, as defined in RFC 3246 [28]. The expedited forwarding PHB is identified based on the following parameters:

- Ensures a minimum departure rate to provide the lowest possible delay to delay-sensitive applications;
- Guarantees bandwidth to prevent starvation of the application if there are multiple applications using expedited forwarding PHB;
- Polices bandwidth to prevent starvation of other applications or classes that are not using this PHB.

The EF PHB is characterized with low delay, low packet loss and low jitter and is suitable for voice and other real-time services. Since an overload of EF traffic will cause queuing delays and affect the jitter and delay tolerances, a strict priority queuing is in most cases implemented for this class. The recommended DSCP binary value for packets requiring expedited forwarding is 101110 (46 or 0 × 2E in decimal and hexadecimal system, respectively). Non-DS compliant devices will regard EF DSCP value as IP precedence 5 (101), which is the highest user-definable IP precedence, and is typically used for delay-sensitive traffic such as voice over IP as shown in Fig. 2.10b.

C. *Assured forwarding (AF) PHB*

The IETF defines the assured forwarding behavior in RFC 2597 [29] and RFC 3260 [30]. This class gives assurance of delivery under defined conditions, i.e., provides assurance of delivery as long as the traffic does not exceed a certain subscribed rate. Traffic that exceeds the subscription rate faces a higher probability of being dropped if congestion occurs. Different multimedia applications, which are characterized by variable bit rate and are sensitive to packet loss, are classified as AF PHB.

The assured forwarding PHB is identified based on the following parameters:

- Guarantees a certain amount of bandwidth to an AF class;
- Allows access to extra bandwidth, if available;
- Packets requiring AF PHB should be marked with DSCP value 'aaaddO' where 'aaa' is a binary value of the class and 'dd' is the drop probability (Fig. 2.10c).

As a result, and in contrast to the two previous classes, the standard defines four separate AF classes within the AF group (AF1, AF2, AF3, and AF4). Within each class, packets are given drop precedence (high, medium, or low), i.e., within each class, the standard now defines three subclasses with a different drop probability. For example, within the AF1 class, the standard now defines three different subclasses AF11, AF12, and AF13, each with a different drop probability as shown in Fig. 2.10c The combination of classes and drop precedence gives twelve separate DSCP encodings, from AF11 to AF43 as shown in Table 2.3.

In case of congestion between classes, the traffic in the higher class is given priority. For example, AF4 class has higher priority than class AF3. If congestion occurs within a class, the packets with the higher drop precedence are discarded first. For example, the probability of dropping the packets that belong to the subclass AF33 is greater than the probability of dropping packets belonging to the

Table 2.3 AF PHB definition

PHB	Class selector	Drop probability	DS code point	DSCP bit (binary)
AF	1	1: Low	AF11	001 010 (10)
		2: Medium	AF12	001 100 (12)
		3: High	AF13	001 110 (14)
AF	2	1: Low	AF21	010 010 (18)
		2: Medium	AF22	010 100 (20)
		3: High	AF23	010 110 (22)
AF	3	1: Low	AF31	011 010 (26)
		2: Medium	AF32	011 100 (28)
		3: High	AF33	011 110(30)
AF	4	1: Low	AF41	100 010 (34)
		2: Medium	AF42	100 100 (36)
		3: High	AF43	100 110 (38)

subclass AF31, i.e., the priority of a given AF33 subclass packet is lower in comparison with the packet that belongs to the AF31 subclass as shown in Fig. 2.10c.

To allow the AF PHB to be used in many different operating environments, the dropping algorithm control parameters must be independently configured for each defined packet drop precedence value and for each AF class. In practice, this specification allows implementation of various configurations that define policies for packet discard. Moreover, an AF implementation must attempt to minimize long-term congestion within each class, while allowing short-term congestion (usually resulting from bursts). Accordingly, in practice, AF class uses more balanced queue servicing algorithms than , such as fair queuing or weighted fair queuing.

D. *Class Selector PHBs*

DiffServ defines the class selector PHB in order to maintain backward compatibility with network devices that are still using the precedence field. The class selector code points are of the form 'xxx000' as shown in Fig. 2.10d. The first three bits are the IP precedence bits. Each IP precedence value can be mapped into a DiffServ class. If a packet is received from a non-DiffServ-aware router that used IP precedence markings, the DiffServ router can still understand the encoding as a class selector code point. Table 2.4 summarizes DSCP specification and backward compatibility with the IP precedence definition.

As we have previously explained, the QoS mechanism specifies a guaranteed network data performance level, i.e., the mechanism which specifies the implementation of QoS includes the setup of the QoS parameters (bandwidth, average packet delay, jitter, and packet loss) in order to improve network performance and meet the demands of end-users. QoS guarantees are important if the network capacity is insufficient, especially for real-time streaming multimedia applications,

Table 2.4 DSCP specification

IP precedence (3 bits)			DSCP (6 bits)				
Name	Value	Bits	PHB	Class selector	Drop probability	DS code point	DSCP bit (binary)
Routine	0	000	Default				
Priority	1	001	AF	1	1: Low	AF11	001 010 (10)
					2: Medium	AF12	001 100 (12)
					3: High	AF13	001 110 (14)
Immediate	2	010	AF	2	1: Low	AF21	010 010 (18)
					2: Medium	AF22	010 100 (20)
					3: High	AF23	010 110 (22)
Flash	3	011	AF	3	1: Low	AF31	011 010 (26)
					2: Medium	AF32	011 100 (28)
					3: High	AF33	011 110 (30)
Flash override	4	100	AF	4	1: Low	AF41	100 010 (34)
					2: Medium	AF42	100 100 (36)
					3: High	AF43	100 110 (38)
Critical	5	101	EF			EF	101 110 (46)
Inter-network control	6	110					(48–55)
Network control	7	111					(56–63)

since they often require a fixed bit rate and are delay sensitive. In practice, for the successful transmission of voice, video applications, and data, QoS parameters' value must be within the defined ranges.

Voice applications have smooth transmission characteristics, and these applications are sensitive to delay, delay variation, and packet loss. Voice applications demand the following values of QoS parameters (one-way requirements):

- Latency lower than 150 ms;
- Jitter lower than 30 ms;
- Loss lower than 1%;
- Voice transmission requires guaranteed priority bandwidth per call. In practice, QoS implementation must provide 150 bps (+ Layer 2 overhead) guaranteed bandwidth for voice-control traffic per call.

In contrast to voice applications, video applications have bursty transmission characteristics but they are also sensitive to delay and packet loss rate. Video applications demand the following values of QoS parameters:

- Latency lower than 150 ms;
- Jitter lower than 30 ms;
- Loss lower than 1%;
- Minimum priority bandwidth guarantee required is as follows:

- Video stream plus 20% of required bandwidth;
- For example, a 384-kbps stream would require 460 kbps of bandwidth.

Data transmission requirements:

- Different applications have different traffic characteristics;
- Different versions of the same application can have different traffic characteristics.

Based on the previous analyses, one of the most frequently used QoS policies in a multiservice network defines the following traffic classes:

- Voice traffic—Low latency, low jitter, and low packet loss;
- Mission critical—Guaranteed latency and guaranteed delivery;
- Transactional traffic—Guaranteed traffic;
- Best-effort traffic—No delivery guarantee.

Different voice, video, and data applications are further classified and marked to belong to EF class, AF subclasses, and BE class. An example of the possible classification and marking in service provider's network is presented in Fig. 2.11.

Application	L3 Classification			L2		SP Service Classes
	IPP	PHB	DSCP	CoS		
Routing	6	CS6	48	6	EF	
Voice	5	EF	46	5		Controlled Latency 35%
Videoconferencing	5	AF41	34	4	AF41 → CS5	
Streaming video	4	AF42	36	4		
Mission-critical data	3	AF31	26	3	CS6	
Call-signaling	3	AF32	28	3	AF31	Controlled Load 1 43%
Transactional data	2	AF21	18	2		
Network management	2	CS2	16	2	AF21→CS3	
Bulk data	1	AF11	10	1	AF11 → AF21	Controlled Load 2 15%
Scavenger	1	CS1	8	1		
Best effort	0	0	0	0		Best Effort 7%

Fig. 2.11 Example of DSCP implementation

2.4   QoS Mechanisms

QoS mechanisms represent a set of techniques or tools which allow traffic classification in an IP network based on models of differentiated services and enable a
given family of basic parameters of QoS (bandwidth, delay, delay variation, and
packet loss percentage) to be maintained in pre-defined boundaries on a source to
destination packets' route. QoS mechanisms can be divided into three main categories (Fig. 2.12):

A. Classification;
B. Scheduling;
C. Network provisioning.

2.4.1   Classification

As previously explained, the lowest service level that a network can provide is the
best-effort service, which does not provide QoS support. Within the best-effort
service, all traffic is handled equally regardless of the application or host that
generated the traffic. However, some applications need QoS support, requiring
better than best-effort service, such as differentiated or guaranteed service.

For a network to provide selective services to certain applications, first of all, the
network requires a classification mechanism that can differentiate between different
applications. The classification mechanism identifies and separates different traffic
into flows or groups of flows (aggregated flows or classes) for identifying a behavior aggregate to which a packet belongs. Therefore, each flow or each aggregated flow can be handled selectively.

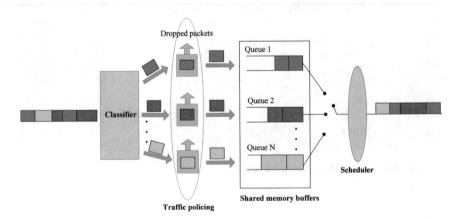

Fig. 2.12 QoS mechanisms

As explained above, packages can be classified either within the second layer of the OSI (Open Systems Interconnection) model using CoS bits (defined in the tag field of the Ethernet frame header [31]), or within the third layer of the OSI model using IP precedence or DSCP bits (defined in the ToS filed of the IP packet header) as shown in Fig. 2.13. The classification of packages should be made as close as possible to the source package so that they would have appropriate treatment in the path through the network. The classification mechanism can be implemented in different network devices (end hosts, intermediate devices such as switches, routers, access points). Figure 2.14 shows a simplified diagram of a classification module that resides on the end host and on the intermediated device.

Application traffic (at the end host), or incoming traffic from other hosts (at the intermediate device), is identified by the classification mechanism and is forwarded to the appropriate queue awaiting service from other mechanisms, for example, the packet scheduler. The granularity level of the classification mechanism can be per-user, per-flow, or per-class depending on the type of QoS services provided. For example, per-flow QoS service requires per-flow classification while per-class QoS service requires per-class classification. To identify and classify the traffic, the traffic classification mechanism requires some type of tagging or marking of the packets. There are a number of traffic classification approaches. Some of approaches are suitable for end hosts and some for intermediate hosts. Figure 2.13 shows an example of some traffic classification approaches which are implemented in the different OSI layers.

Most QoS mechanisms include some type of classification, but a small number of mechanisms also include marking capability as shown in Fig. 2.15.

OSI model	TCP/IP model	Classification Techniques
Application layer	Application layer	User/Application identification
Presentation layer		
Session layer		
Transport layer	Transport layer	Flow
Network layer	Internet layer	IP ToS, DSCP
Data-link layer	Network access layer	802.1P/Q Classification
Physical layer		

Fig. 2.13 Example of the existing classification on each layer of OSI and TCP/IP model

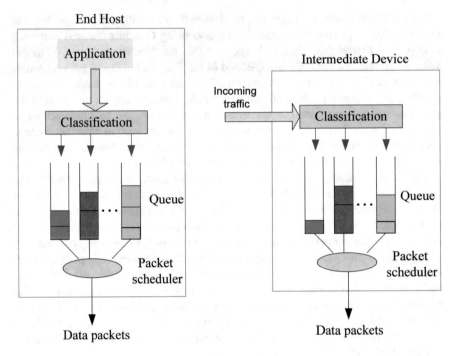

Fig. 2.14 Implementation of classification in network devices

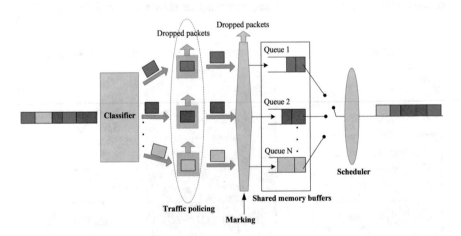

Fig. 2.15 Classification and marking

Marking is a term used for coloring packets by applying a class-identifying value to one of the following markers: IP precedence, DSCP, QoS group (value is local to a router), MPLS experimental bits (can be used only in MPLS-enabled networks), ATM CLP bit (value can be used only within ATM networks), Frame Relay DE bit (value can be used only within Frame Relay networks), or IEEE 802.1Q priority bits (value can be used within LAN-switched networks) [31]. In practice, marking is used after classification and traffic policing. Traffic policing configuration has to determine whether the packet is in a pre-configured profile or out of the profile. Based on that information and the configured parameters, marking configuration must determine whether to pass through, mark down (change the DSCP, or CoS, or any of the previously mentioned parameters), or drop the packet. Furthermore, packets are queued based on the pre-configured (marked) parameters and scheduled for transmission.

2.4.2   Scheduling Mechanism

Packet scheduling is the mechanism that selects a packet for transmission from the packets waiting in the transmission queue. Scheduling mechanism decides which packet from which queue and station are scheduled for transmission in a certain period of time. Packet scheduling controls bandwidth allocation to stations, classes, and applications.

As shown in Fig. 2.16, there are two levels of packet scheduling mechanisms:

- Intra-station packet scheduling: the packet scheduling mechanism that retrieves a packet from a queue within the same host;
- Inter-station packet scheduling: the packet scheduling mechanism that retrieves a packet from a queue from different hosts.

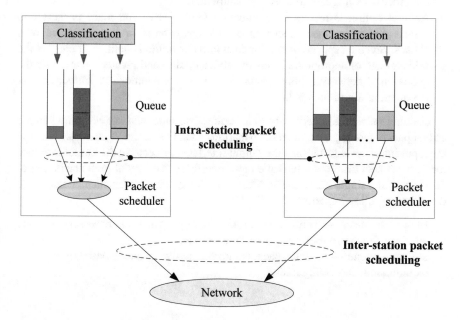

Fig. 2.16 Packet scheduling mechanism

In the intra-station packet scheduling, the status of each queue in a station is known, and hence, the intra-station packet scheduling mechanism is identical to a queuing mechanism. On the contrary, the inter-station packet scheduling mechanism is different from a queuing mechanism because queues are distributed among hosts and there is no central knowledge of the status of each queue. Therefore, inter-station packet scheduling mechanisms require a signaling procedure to coordinate the scheduling among hosts and resource allocation. Because of the similarities between packet scheduling and queuing mechanisms, we further introduce four queuing schemes: FIFO (First In, First Out), SPQ (Strict Priority Queuing), CQ (Custom Queuing), and WFQ (Weight Fair Queuing) and discuss how they support QoS services.

A. *Queuing mechanism*

Transmission buffers in network switches and routers tend to fill rapidly in high-speed networks, and therefore buffering, not bandwidth, represents the key QoS issue. The overloaded buffers cause packet drops, which in turn causes voice or video clipping and skipping. Every point in the network where there is a router or a switch may be a source of transmission or buffering challenges potentially giving rise to poor QoS.

There are several approaches to solving a problem of insufficient bandwidth:

- The best approach is to increase link capacity to accommodate all applications and users with some extra bandwidth to spare. This solution sounds simple enough, but in practice, it brings high costs in terms of money and time it takes to be implemented. Oftentimes, there are also technological limitations related to the process of upgrade to a higher bandwidth;
- Another option is to classify traffic into QoS classes and place packets into queues which are then processed on a priority basis or in accordance with classes. Since a separate queue for each class of traffic is defined, it enables the achievement of appropriate levels of service quality and greater control over the operation of networks. Since this is a cost-effective solution, we further discuss the queuing algorithms below.

Queuing on routes is necessary to accommodate burst and congestion when the arrival packet rate is greater than the departure rate. This is usually the case because the input interface is faster than the output interface, or because the output interface receives packets from other multiple input interfaces. The initial implementation of queuing used a single FIFO queue. More complex queuing mechanism includes the definition of following terms (Fig. 2.17):

- Hardware queue—always uses FIFO strategy, which is necessary for the interface drivers to transmit packet one by one;
- Software queue—schedules packets into the hardware queue based on the QoS requirements and configuration.

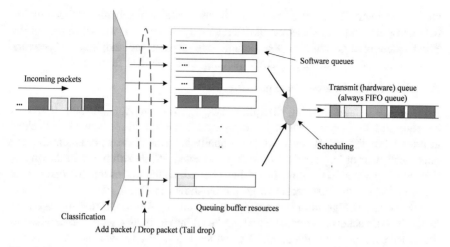

Fig. 2.17 Queuing implementation

Generally, a full hardware queue indicates interface congestion and software queuing is utilized to manage it. When a packet is being forwarded, the router will bypass the software queue if the hardware queue has space in it (no congestion). The implementation of software queuing is optimized for periods when the interface is not congested.

Most queuing mechanisms include the classification of packets. After a packet is classified, a router has to determine whether it can put the packet into the queue or it has to drop the packet. Most queuing mechanisms will drop a packet only if the corresponding queue is full. This is called tail drop. Some mechanisms use a more intelligent dropping scheme, such as random early detection (RED) [20]. If the packet is allowed to be queued, it will be put into the FIFO queue for that class or flow. Packets are than taken from the individual per-class queues and put in the hardware queue.

In practice, to enable queuing, traffic has to be classified according to importance, i.e., business-critical traffic should get enough bandwidth, voice should get enough bandwidth and prioritized forwarding, and the least important traffic should get the remaining bandwidth. Based on each data packet's classification, the packet is placed into the appropriate transmission queue. Time critical voice and video data are classified in such a way that they are placed into a delay- and drop-sensitive queue. Separate queues allow time critical data, such as audio and video, to be transmitted in a priority fashion. Queuing gives the delay-sensitive voice and video data a higher priority in the network switch or router ensuring that the voice or video packet will be transmitted in a timely manner.

In case of network congestion, the queuing mechanism avoids the unnecessary dropping of packets by storing them for a while, hoping that congestion will be alleviated, and that they can be then dispatched. When congestion is heavy and the queue overflows, new arriving packets have to be dropped since there is no other choice for them. There is a wide variety of available mechanisms that provide bandwidth guarantees but, as we have previously mentioned, we will focus on the FIFO queuing algorithm as the default one among the majority of network equipment, priority queuing, custom queuing, and weighted fair queuing.

B. *FIFO (First In, First Out) queuing*

FIFO is the default queuing discipline used by most vendors. In fact, it has its benefits and limitations, but it is still very simple for implementation. Packets arriving from different flows are treated equally in terms of their placement into the queue and their order of arrival is strictly respected. While still being in the queue, they are dispatched in the same order they entered, which means that the first packet that comes in is the first packet to go out as shown in Fig. 2.18.

FIFO queuing algorithm is the simplest one among the congestion management methods. All packets are treated equally, placed into a single queue, and serviced in the order they were received. FIFO provides best-effort service, i.e., it does not provide service differentiation in terms of bandwidth and delay. The high bandwidth flows will get a larger bandwidth portion in comparison with the low bandwidth flows. Aggressive flows send a large number of packets which occupy the majority of bandwidth. Time-sensitive flows send a modest number of packets, and most are dropped due to the queue always being full of aggressive flow packets. It is possible to improve QoS support by adding a traffic policing mechanism, for limiting the rate of each flow, and by implementing admission control. In practice, the size of queue affects delay, jitter, and packet loss. Namely, the queue has a finite

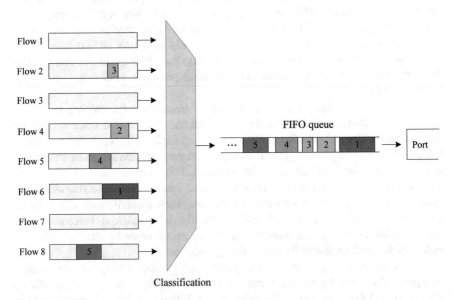

Fig. 2.18 FIFO queuing

size; hence, in the case of congestion, the packet that needs to be added to the queue could be dropped (tail drop mechanism). To overcome this problem, the queue could be lengthened, which decreases the probability of tail drop. With a longer queue, however, the average delay and jitter increases, because packet may be queued behind a large number of other packets.

FIFO queuing offers the following benefits:

- FIFO queuing places an extremely low load on the system in comparison with other queuing mechanisms;
- FIFO queuing is predictable, and delay is determined by the maximum length of the queue;
- FIFO queuing does not add significant queuing delay at each hop as long as the queue length remains low.

FIFO queuing also poses the following limitations:

- FIFO queuing is extremely unfair when an aggressive flow contests with a time-sensitive flow. Aggressive flows send a large number of packets, many of which are dropped. Time-sensitive flows send a modest number of packets, and most are dropped due to the queue always being full of aggressive flow packets. This behavior is called starvation;
- When the queue is full, packets entering the queue have to wait longer. When the queue is not full, packets entering the queue do not have to wait until;
- A single FIFO queue does not allow routers to organize buffered packets and service one class of traffic differently from other classes of traffic. Therefore, FIFO queuing can result in increased delay, jitter, and a reduction in the amount of output bandwidth consumed by TCP applications.

C. *Priority Queuing*

Priority packet scheduling schedules packets based on the assigned priority order. Packets in higher priority queues are always transmitted before packets in lower priority queues. A lower priority queue has a chance to transmit packets only when there are no packets waiting in a higher priority queue.

As shown in Fig. 2.19, a scheduler first empties the higher priority queue, next the medium one, and finally the low priority one, i.e., as soon as there are packets in the high priority queue, it must be served first until it becomes empty, then the medium priority queue and finally the low priority one. SPQ provides differentiated services in both bandwidth and delay. The highest priority queue always receives bandwidth (up to the total bandwidth), and the lower priority queues receive the remaining bandwidth. Therefore, higher priority queues always experience lower delay than the lower priority queues. The aggressive use of bandwidth by the high priority queues can starve the low priority queues. As it was the case with the FIFO algorithm, the QoS support can be improved by traffic policing , which limits the rate of each flow, and by the implementation of the admission control mechanism.

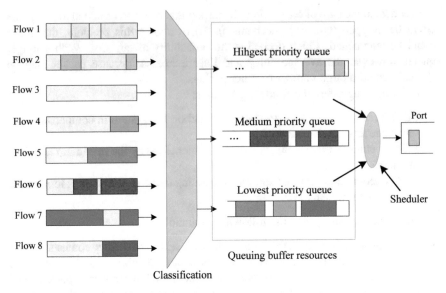

Fig. 2.19 Implementation of priority queuing

D. *Custom Queuing*

CQ is one of the most popular queuing strategies. CQ was originally implemented to address the clear shortcomings of PQ. It enables us to configure how many queues are to be used, what applications will use which queues, and how the queues will be serviced. While PQ has only four queues, CQ is composed of 17 queues, numbered from 0 to 16. Queue 0 is a system queue reserved for the control traffic, and hence, application data cannot starve the critical network control traffic. Consequently, this queue cannot be set by the user and the remaining queues, queues 1–16, are the settable queues.

Custom queuing mechanism is based on the weighted round-robin (WRR) approach [20]. In this approach, packets are accessed in the round-robin style but queues can be given priorities called 'weights.' The idea behinds CQ is the assignment of the percentage of the available bandwidth to each of the 16 queues until the 100% of total availability is reached and this is achieved by the application of the 'weight' definition. Namely, the WRR approach places a certain weight on various queues to service a different number of bytes or packets from the queues during a round-robin cycle. WRR mechanism implemented in the custom queuing takes a certain predetermined amount of bytes defined by the weight factor from each queue during each pass as shown in Fig. 2.20. In practice, weight factors, i.e., number of bytes, can be defined and this allows administrators to approximately specify how much of the bandwidth each queue will receive.

However, although more efficient than the previously described schemes, there are still three main problems that exist in custom queuing.

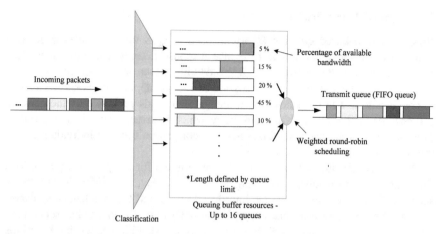

Fig. 2.20 Implementation of custom queuing

The first problem of custom queuing arises from the fact that weight of the queue is configured in bytes, and therefore, the accuracy of the WRR depends on the weight (byte count) and the MTU (Maximum Transmission Unit). Namely, the defined percentage shares cannot be used to set the queues, and instead, the 'byte count,' the value of which is dependent on flow's packet size in the queue, must be used. If the ratio between the byte count and the MTU is too small, the WRR queuing will not allocate bandwidth accurately. On the other hand, if the ratio is too large, WRR will cause long delays. Since queues may have different average packet sizes (e.g.,. voice packets of 60 bytes and TCP packets of 1500 bytes), this may lead to the undesirable bandwidth distribution ratio. For example, if queue byte counter is 100 bytes and a packet of 1500 bytes is in the queue, the packet will still be sent anyway since the counter is nonzero. As a result, the queue will receive fourteen times more bandwidth (1400 bytes more instead of 100 bytes) in this round than it should have received. This is the consequence of the limitation of the round-robin scheduling that cannot take less than one packet from the each queue. In order to make the distribution fair, every queue's byte counter should be proportional to the queue's average packet size, which is, in practice, very difficult to achieve.

The second problem is the classification, i.e., the implementation of the CQ scheme. The classification can be based on different parameters (protocol, source interface, source and destination addresses, and ports) where the control by address demands the device configuration be set in advance.

The last problem arises from the fact that in CQ, the traffic within each queue competes directly with all other traffic in the same queue. So, for example, if one user sends a burst of application traffic that fills one of the queues, this will cause tail drops for other users whose traffic occupies the same queue. This will cause a smaller version of the global problem of a FIFO queue.

E. *Weighted Fair Queuing*

WFQ is a flow-based queuing algorithm used in QoS that schedules interactive traffic to the front of the queue to reduce response time, and it fairly shares the remaining bandwidth between high bandwidth flows as shown in Fig. 2.21. A stream of packets within a single session of a single application is known as flow or conversation. WFQ is a flow-based method that sends packets over the network and ensures packet transmission efficiency which is critical for interactive traffic. This method automatically stabilizes network congestion between individual packet transmission flows.

WFQ was introduced as a solution to the problem for the previously described FIFO queuing mechanism, including starvation, delay, and jitter. Moreover, it successfully resolves the problems introduced in PQ and CQ mechanisms, starvation of other lower priority classes, and long delays among others. For scheduling purposes, in WFQ, the length of the queue is not measured in packets but by means of the time it would take to transmit all the packets in the queue. WFQ adapts the number of flows and allocates equal amounts of bandwidth to each flow. Small packet flows which are usually interactive flows (voice and video) typically receive better service because they do not need a lot of bandwidth. Weight can be determined as the required QoS (IP precedence, RSVP), or inversely proportional flow throughput. For example, a link data rate of R, with N active data flows with weights $w_1, w_2, \ldots w_N$ and data flow number i, will achieve an average data rate of

$$R_i = \frac{Rw_i}{(w_1 + w_2 + \ldots + w_N)} \qquad (2.1)$$

The weight factors can be calculated from the IP precedence value. The result is that the router gives flows with higher IP precedence values a larger share of the

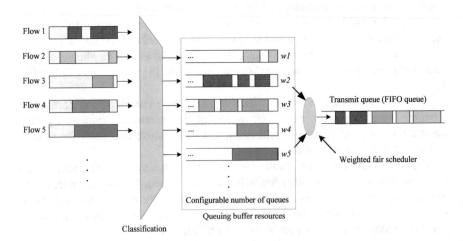

Fig. 2.21 Implementation of weighted fair queuing

bandwidth than those with lower precedence. For example, a flow with flash override IP precedence will get five times the bandwidth of a packet with routine IP precedence. After calculation, a weighting factor is assigned to each competing connection prior to data transmission. The weighting factor is chosen proportionally to the connection's bandwidth share. The WRR scheduler serves all connections in a cyclic way (round-robin) where each connection is allowed to send only a number of packets according to its weight factor.

These fair queueing algorithms tend to do three things. First, they prevent individual flows from interfering with one another. Second, they tend to reduce queueing latency for applications with smaller packets. And third, they ensure that all of the packets from a given flow are delivered in the same order they were initially sent. It is a fact that, when using a network with WFQ, an end-to-end delay bound can be guaranteed. By regulating the WFQ weights dynamically, WFQ can be utilized for controlling QoS, for example, to achieve guaranteed data rate.

Although bandwidth is allocated fairly among all flows, unfairness is reinstated by introducing weight (IP precedence) to give proportionately more bandwidth to flows with higher weights. WFQ has to classify individual flows using the following information taken from the IP/TCP/UDP headers. These parameters are used as input for a hash algorithm that produces a fixed length number that is used as the index of the queue:

- Source IP address;
- Destination IP address;
- Protocol number to identify TCP or UDP;
- Type of service field;
- Source TCP/UDP port number;
- Destination TCP/UDP port number.

The main benefits of WFQ are as follows:

- Simple configuration (classification does not have to be configured);
- Guarantees throughput to all flows;
- Dedicated queues for each flow (referred to as conversations), messages are sorted into conversations reducing starvation, delay, and jitter within the queue;
- Allocating bandwidth fairly and accurately among all flows, reducing scheduling delay and guaranteeing service;
- Drops packets of most aggressive flows;
- Supported on most platforms.

The main drawbacks are as follows:

- All drawbacks of FIFO queuing within a single queue;
- Multiple flows can end up in one queue;
- Does not support the configuration of classification;
- Cannot provide fixed bandwidth guarantees;
- Performance limitations due to complex classification and scheduling mechanisms.

2.4.3   Network Provisioning

Networking provisioning requires a thorough network analysis to determine parameters for services that are being deployed in the network. The result of monitoring techniques and provisioning is the allocation of bandwidth among all classes in times of congestion. Services are implemented by defining PHB properties. PHBs are implemented by using the available QoS mechanisms in networks devices.

The main benefits of provisioning are as follows:

- QoS does not create bandwidth;
- QoS manages bandwidth usage among multiple classes;
- QoS gives better service to a well-provisioned class with respect to another class.

In order to be effective in providing network QoS, it is necessary to affect the provisioning and configuration of the traffic handling mechanisms consistently, across multiple network devices. Provisioning and configuration mechanisms include the following:

- RSVP signaling;
- Policy mechanisms and protocols;
- Management tools and protocols.

Chapter 3
PON Evolution

3.1 PON Development

The growing popularity of the Internet and different multimedia applications (IPTV, VoD, HDTV and many others) as well as a constant increase in the number of users are the key factors behind the development of new access technologies which are highly likely to meet new bandwidth requirements. As we have emphasized in the introductory section, the access network based on copper has distance and bandwidth limitations and does not have enough capacity and potential for the future development of applications and networks as shown in Fig. 3.1.

In such a situation, fiber optical access networks of today are getting increasingly more attention as they offer the ultimate solution for delivering different services to end-users. Due to the lack of active units in the light path, the architecture of PON is simple, cost-effective, and offers bandwidth which is highly unlikely to be achieved by other access methods. In the same time, constant development of semiconductor lasers [32–33] greatly increases the efficiency of optical transmitters in central unit and overall transmission efficiency as well. Moreover, besides optical transmitters, the development of various optical receivers, especially fast semiconductor photodetectors [34–36] and new cheap polymer photodetectors [37], further improves the performances of optical units and therefore the performances of the whole system. In the future, one of the more prominent issues will be low price of building optoelectronic elements within optical units, what may be attained by integrating all necessary optical and electronic component on scalable chip using low-cost silicon waveguides [38–40] in CMOS technology. This means that we may expect in near future sophisticated, low cost with high bandwidth, adaptable, multiservice optical access networks.

PON advantages include (Table 3.1) [41]

© Academic Mind and Springer International Publishing AG 2017
M. Radivojević and P. Matavulj, *The Emerging WDM EPON*,
DOI 10.1007/978-3-319-54224-9_3

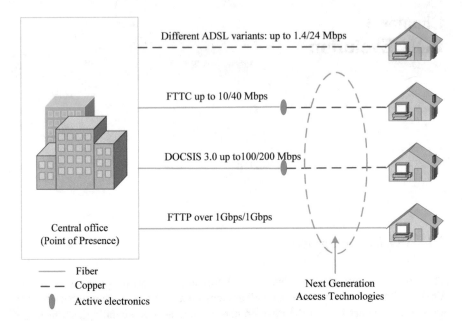

Fig. 3.1 Comparison of access technologies. *Note* Maximum download/upload speed

- Very high-speed data transmission of up to 1 Gbps to home and business due to the deeper fiber penetration;
- Maintenance cost reduction due to the lack of electronics between CO and customers—PONs eliminate the necessity of installing active multiplexers at splitting locations, thus relieving network operators of the gruesome task of maintaining active curbside units and providing power to them;
- Instead of active devices in these locations, PONs use small passive optical splitters, located in splice trays and deployed as part of the optical fiber cable plant;

Table 3.1 Technology comparison

Service	Medium	Mbps down	Mbps up	Max. distance at full rate (km)	Standards
ADSL	Twisted pair	8	1	2.4	ITU G.992.1
VDSL	Twisted pair	50	6.4	0.4	ITU G.993.1
ADSL2+	Twisted pair	24	1.0	1.5	ITU G.992.5
HFC	Coax	55	30	25	DOCSIS 2.0
HFC	Coax	200	100	20	DOCSIS 3.0
BPON	Fiber	155/622	155	20	ITU G.983
GPON	Fiber	1244/2488	155/622/1244/2488	20	ITU G.984
EPON	Fiber	1250	1250	20	IEEE 802.3ah

- PONs minimize fiber deployment in both the local exchange office and the local loop—low-cost solution due to fiber and CO interface shared by several customers;
- Constant data rate regardless of reach—PONs allow long reach between central offices and customer premises, operating at distances over 20 km;
- Multiple applications including data (IP), video, and voice (triple-play)—operating in the downstream as a broadcast network, PONs allow video broadcasting as either IP video or analog video using a separate wavelength overlay;
- Being optically transparent end to end, PONs allow upgrades to higher bit rates by adding more wavelengths.

The passive nature of the PON medium and the fact that the electronics is only at the ends means that provisioning and repair are accomplished much more quickly than with systems with active equipment. Passive optical networks, even though based on a shared medium, possess at least ten times the capacity at usable subscriber distances and require no electronics in comparison with DSL and cable technologies.

Although PONs can exist in three basic configurations (tree, bus, and ring), the tree topology is favored due to its smaller variation in the signal power from different end stations as shown in Fig. 3.2. In the tree topology, a passive optical network is considered as a point-to-multipoint fiber to the premises network architecture, in which passive unpowered optical splitter is used to enable a single optical fiber to serve multiple premises as shown in Fig. 3.3.

A PON configuration reduces the amount of fiber and CO equipment required compared with point-to-point architectures. As explained in Sect. 1.1.4, a PON consists of an OLT at the service provider's central office and a number of ONUs at the subscriber's end. Besides an ONU term, the optoelectronic element at the user side is sometimes referred as the network interface device (NID) as well as the optical network terminal (ONT). The portion of the network between OLT and ONUs consists of fibers and passive optical splitter that is defined as optical distribution network (ODN).

The OLT is the main element of the network and is usually placed in the CO. It is a network element with PON line card, basically an aggregation switch that serves as the point of origination for fiber to the premises transmissions coming into and out of the optical access network. This unit works as an interface between the core network and PON.

The OLT provides the interface between the PON and the service provider's network that typically includes

- Internet Protocol (IP) traffic over 100 Mbps;
- Ethernet 1 or 10 Gbps;
- Standard time-division-multiplexed (TDM) interfaces such as SONET (Synchronous Optical Network) or SDH (Synchronous Digital Hierarchy);
- ATM (Asynchronous Transfer Mode) at 155−22 Mbps.

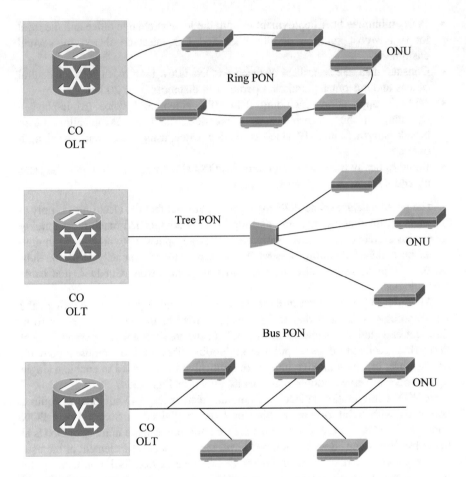

Fig. 3.2 PON topologies

An ONU is a device that terminates PON and presents customer service inter-faces to the user and at the same time serves as an interface to the network. It provides several interfaces for accessing triple-play services, and in the upper side, it connects with the OLT via optical splitter. In practice, some forms of imple-mentation demand, besides ONU, the implementation of a separate subscriber unit to provide services such as telephony, Ethernet data, or video.

In multiple tenant units, the ONU may be bridged to a customer premise device within the individual dwelling unit using technologies such as Ethernet over twisted pair, coax, or DSL. ONT is only a term that is used to describe the situation in which an ONU connects a single end-user to the network. Besides the ONU and ONT terms, a PON employs a passive splitter/combiner to split the optical signal from one fiber into several fibers and, reciprocally, to combine optical signals from multiple fibers into one. Splitters can be placed anywhere between the CO and

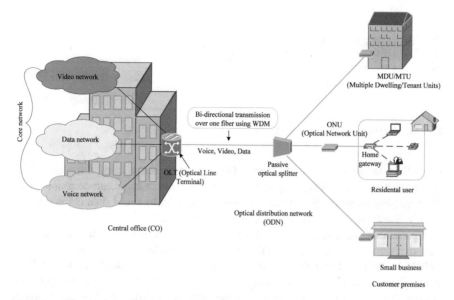

Fig. 3.3 Realization of multiservice PON

subscriber premises. It is used to connect an optical port of OLT with multiple subscribers. The directional properties of a passive splitter/combiner are such that an ONU transmission cannot be detected by other ONUs. Passive splitters/ combiners provide complete path transparency.

The communication path from the OLT to the ONU is referred to as downstream, whereas the reverse path is referred to as upstream. The downstream and upstream signals are carried over the same fiber. In the downstream direction (from the OLT to ONUs), a PON is a point-to-multipoint network. The OLT typically has the entire downstream bandwidth available to it at all times. In the upstream direction, a PON is a multipoint-to-point network where multiple ONUs transmit toward the OLT. A PON takes advantage of wavelength division multiplexing (WDM), using one wavelength for downstream traffic and another for upstream traffic on a single non-dispersion-shifted fiber (ITU-T G.652). PON and its various flavors, such as APON (ATM PON), GPON (Gigabit PON), and EPON (Ethernet PON), have the same basic wavelength plan and use the 1490 nm wavelength for downstream traffic and 1310 nm wavelength for upstream traffic as shown in Fig. 3.4. 1550 nm is reserved for optional overlay services, typically RF (analog) video. Signals are inserted or extracted from the fiber using a coarse wavelength division multiplexing (CWDM) filters at the CO and subscriber premises.

Since PON is a shared network, the OLT sends a single stream of downstream traffic that is seen by all ONUs. Each ONU only reads the content of those packets that are addressed to it. Encryption is used to prevent eavesdropping on downstream traffic. However, in the upstream direction, the situation is quite different. Although directional properties of a passive splitter/combiner do not allow the transmission of

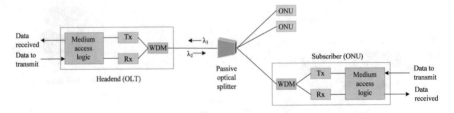

Fig. 3.4 Architecture of OLT and ONUs

one ONU to be detected by other ONUs, data streams from different ONUs transmitted simultaneously may still collide. Thus, in the upstream direction, PON must employ a certain channel separation mechanism to avoid data collisions and evenly share the trunk channel capacity and resources.

Upstream signals are combined by means of a multiple access protocol, namely time-division multiple access (TDMA) or wavelength division multiple access (WDMA). Today, TDM PON is widely implemented in the field, but WDM PON gains more attention as a result of the constant development of services, and consequently, we analyze this option as the optimal candidate for the realization of NGN PONs.

In the following sections, we first discuss TDM PON networks, i.e., PON flavors such as APON, GPON, and EPON. We focus on the implementation and operation of EPONs as the cost-effective and, in our belief, the optimal solution for the realization of optical access networks. The special attention is given to the QOS support as a key factor for the realization of a truly broadband infrastructure. Moreover, even though the TDM approach is widely implemented in the field, WDM approach today gains more attention with the rapid Internet traffic increase. Accordingly, we further present and analyze various models for the realization of hybrid TDM/WDM EPONs with QoS support as the optimal candidate for the realization of NGN PONs.

3.2   TDM PON

In TDM PONs, OLT allocates a timeslot, or a transmission window, for data transmission to each ONU in the system as shown in Fig. 3.5. ONU buffers packets received from end-users until it receives the allocated timeslot. Upon the allocation of the timeslot, ONU sends out buffered packets (timeslot is capable of carrying several Ethernet frames) at the full transmission rate of the upstream channel.

Since every ONU transmits buffered packets only in the precisely allocated timeslots, it is obvious that data collision is avoided. The allocated timeslot can be:

- Fixed—In fixed timeslot allocation, also called static bandwidth allocation, each ONU is configured to start and stop transmission at the predetermined repeating intervals. Fixed TDMA schemes are simple to implement, but due to the bursty

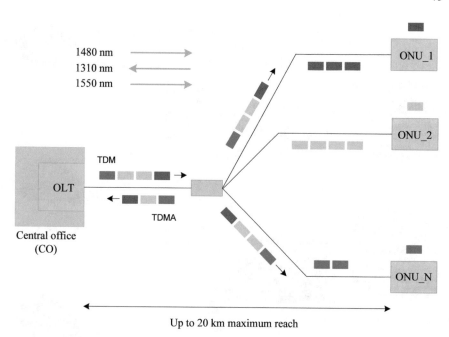

1480 nm →
1310 nm ←
1550 nm →

TDM

OLT

TDMA

Central office
(CO)

ONU_1

ONU_2

ONU_N

Up to 20 km maximum reach

Fig. 3.5 TDM PON characteristics

nature of network traffic, it may result in a situation in which some timeslots overflow even under very light load. As a result, packets will be delayed for several timeslots, even though other timeslots are not fully used. As a result, the upstream bandwidth will be underutilized. For this reason, the static allocation is not a preferred configuration nowadays since the majority of multimedia applications have bursty transmission characteristics;

- Variable—In variable timeslot allocation, also called statistical multiplexing scheme, OLT dynamically allocates a variable timeslot to each ONU based on the instantaneous bandwidth demand (instantaneous queue size) of the ONUs. Today, we may find different allocation schemes, including schemes utilizing traffic priority and QoS, service-level agreements (SLAs), and oversubscription ratios [42]. This approach increases bandwidth utilization and system efficiency in comparison with the fixed timeslot allocation.

One of the major advantages of this approach is that all ONUs can operate on the same wavelength and be absolutely identical. Consequently, an ONU transceiver must operate at the full line rate, although the allocated bandwidth may be lower in the field device. Moreover, since the OLT needs a single receiver, this solution is highly cost-effective.

Today, several alternative architectures for the realization of TDMA PON-based access networks are standardized and used for the realization of optical access networks. The main difference between the currently used PON kinds is the choice of the bearer protocol; hence, we discuss (Fig. 3.6):

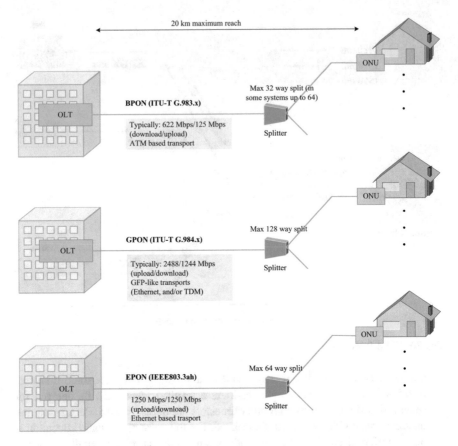

Fig. 3.6 TDM PON architecture and technologies

- ATM-based PON also called Broadband PON (BPON);
- PON utilization generic frame procedure (GFP) or the gigabit PON (GPON);
- Ethernet-based PON (EPON).

3.2.1 ATM PON (APON)

The first PON activity was initiated in the mid-1990s when a group of major network operators established the full service access networks (FSAN) consortium. The four FSAN members (NTT, British Telecom, BellSouth, and France Telecom) issued a common technical specification for ATM (Asynchronous Transfer Mode) subscriber systems. The group understood the need to develop future access networks and realized that the wide benefits of a new approach could be achieved through adopting a common set of specifications. Moreover, the consortium concluded that the deployment of fiber-based broadband networks could be cost-effective if their component parts were produced in large quantities for tens of millions of access lines. This well-intended initiative was taken over by ITU-T, and several products that are standard (G.983) compliant were produced [43]. It established the general requirements for PON system, i.e., APON that uses ATM as the underlying transport mechanism to carry users' data and support at a 155 Mbps transmission rate.

The name BPON (Broadband PON) was introduced since the name APON was confusing and led to the conclusion that only ATM services could be provided to end-users. Changing the name from ATM PON to Broadband PON reflected the fact that BPON systems offer broadband services including Ethernet access, video distribution, and high-speed leased line services. The APON standards were later enhanced to support 622 Mbps transmission rates as well as additional features in the form of protection, dynamic bandwidth allocation (DBA) algorithm, and others.

The main characteristics of BPON include

- 53-byte ATM cells with mini-cells in transmission convergence (TC) layer;
- Downstream 'grants' control the sending of upstream cells;
- Rates up to 622.08 Mbps symmetrical and 1240/622 asymmetrical have been standardized;
- Wavelengths: 1260–1360 nm up, 1480–1580 nm down;
- Transport capability—native ATM, TDM (T1/E1) by circuit emulation, Ethernet by emulation;
- 32-way split (some systems 64 way);
- Support for multicast transmission;
- Standardized in G983.x series in ITU.

In the BPON, the process of transporting data downstream is different from transporting data upstream, as illustrated in Fig. 3.7. In the ODN, simultaneous transmission on the same fiber is enabled by using different wavelengths for each direction: 1490 nm for downstream (from the OLT to the ONUs) and 1310 nm for upstream (from the ONUs to the OLT). In an ATM cell-based PON, the downstream signal is broadcast to all ONUs, as shown in Fig. 3.7a. Each ONU discards or accepts the incoming cells depending on the cell header addressing. Encryption is necessary to maintain privacy, since the downstream signal is broadcast and each ONU receives all the information. In the upstream direction, the system uses a TDMA (time-division multiple access) protocol to accede the bandwidth as shown in Fig. 3.7b. The timeslots are synchronized so that packets traveling upstream from the ONTs do not collide with each other when hopping onto the common fiber. The OLT controls the transmissions from each ONU by sending grants or permissions to them. In order to avoid collisions between transmissions from different ONUs, a technique, ranging, is executed to measure the logical distance between the ONUs and the OLT. As a result, each ONU adjusts its transmission time properly, and thus, the effects of propagation delays are successfully avoided. Moreover, at the

(a) Downstream transmission

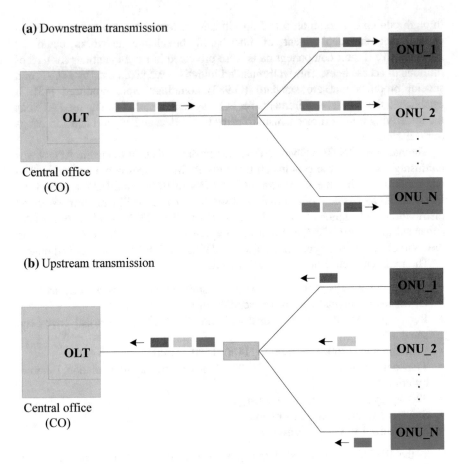

(b) Upstream transmission

Fig. 3.7 a Downstream transmission; **b** Upstream transmission

request of the OLT, an ONT may also report to the OLT its packet-buffer length, and the OLT can subsequently reapportion grants accordingly to accommodate the traffic needs of the ONT.

The transmission protocol is based on a downstream frame of 56 ATM cells (53-bytes each) for the basic rate of 155 Mbps, scaling up with bit rate to 224 cells for 622 Mbps. Downstream transmission is a continuous ATM stream at a bit rate of 155.52 Mbps or 622.08 Mbps with dedicated PLOAM (physical layer operation, administration, and maintenance) cells inserted into the data stream. The PLOAM cell is the center of the PON's operation, in both directions [42]. The downstream frame is constructed from two PLOAM cells, one at the beginning of the frame and one in the middle, and 54 data ATM cells. Each PLOAM cell contains grants for upstream transmission relating to specific cells within the upstream frame (53 grants for the 53 upstream frame cells are mapped into the PLOAM cells) as well as

OAMP (Operation, Administration, Maintenance and Provisioning) messages as shown in Fig. 3.8.

The upstream frame format consists of 53 cells of 56 bytes each (53 bytes of ATM cell plus 3 bytes overhead) for the basic 155 Mbps rate. Upstream transmission consists of either a data cell, containing ATM data in the form of VPs/VCs (virtual paths/virtual circuits), or a PLOAM cell when granted a PLOAM opportunity from the central OLT. Upstream transmission is in the form of bursts of ATM cells, with a 3-byte physical overhead appended to each 53-byte cell in order to allow burst transmission and reception. Two options are available for the use of upstream capacity: static (fixed) allocation and dynamic bandwidth allocation (DBA). It is possible to operate both options concurrently but the dynamic option is preferred. The upstream bandwidth for each ONU is allocated through the implementation of the dynamic bandwidth allocation (DBA) algorithm. OLT sends grants as permissions for each ONU to send one or more cells in succession to form a T-container (T-cont, transmission container).

As previously explained, data is transported via fixed-size packets of 53-octets. Transmitted data must go through an ATM adaptation layer (AAL), where segmented into a 48-octet unit with a five-octet header (ATM cell tax) added, which results in 53-octet packets called an ATM cell. When the cells arrive at the destination, they are then reassembled back into the original traffic. This is called the segmentation and reassembly (SAR) process. SAR process allows ATM to be easily adapted for voice, video, and data services. Moreover, small, fixed-size

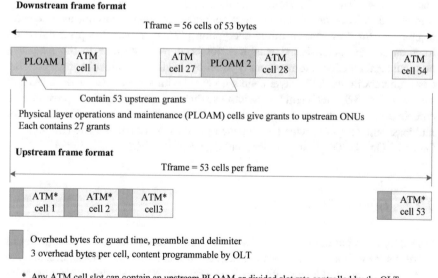

Fig. 3.8 Frame format for 155.52/155.52 Mbps BPON (G.983.1 [43])

packets are optimal for carrying delay-sensitive traffic. On the other hand, the SAR process can be time-consuming even if implemented in hardware. Besides SAR, the provisioning of ATM virtual paths and virtual channels and associated cross-connections also add further complexity. However, because the packets have a fixed size, the upstream traffic control is simplified, allowing the OLT to grant upstream timeslots to individual ONTs so they can send data upstream on the common fiber without collision.

In addition to the previously described functions, the PLOAM cell also has an important role in the process called ranging. As explained, the upstream transmission is based upon TDMA; hence, each ONU can transmit packets upstream only during its assigned timeslot. Ranging is a function which measures the logical distance between each ONU and OLT and decides on the transmission timing when each ONU receives a grant. Namely, the BPON networks ensure collision-free upstream transmission by requiring all ONUs to transmit cells at the equivalent maximum distance of 20 km. In order to discover the equalization delay for each ONU, the OLT sends a ranging grant to an ONU and waits for a response. The round-trip time delay is measured and subtracted from the maximum 40 km round-trip time delay, producing the equalization delay for the ONU. Furthermore, the individual ONUs are configured to insert appropriate delays, the values of which are derived by the ranging process [43]. All ranging-related messages are mapped in the message field of the PLOAM cell. This method is straightforward and effective because of the fixed packet size.

Besides the previously described features and enhancements, the PLOAM cell provides a very rich and exhaustive set of operation, administration and maintenance (OAM) features, including, among others, bit error rate (BER) monitoring, alarms and defects, autodiscovery and automatic ranging, churning as a security mechanism for downstream traffic encryption [42]. However, the price reduction foreseen for the ATM-based gear has not been realized. Due to the extensive protocol segmentation and reassembly to 53-byte cells for protocol conversion, the costs inherent in the ATM layer cannot compete with low-cost Ethernet devices as shown in Fig. 3.9. Inefficient bandwidth use due to fixed cell size, the provisioning complexity, relatively low bit rates defined by the standard along with the lack of multicast support per service have opened the door to Ethernet PON (EPON) and Gigabit PON (GPON)more efficient and cost effective PON technologies.

3.2.2   Gigabit PON (GPON)

Driven by the rapid development of services, the constant increase in the number of end-users, as well as the previously described shortcomings of BPONs, in 2001, the FSAN group initiated a new effort for standardizing PON networks operating at bit rates above 1 Gbps. However, the support for higher bit rate was not the only goal of this initiative. Namely, the overall protocol has been opened for reconsideration in order to achieve the most optimal and efficient solution in terms of support for

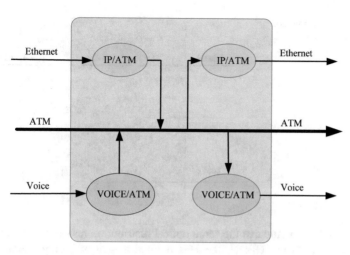

Fig. 3.9 Protocol conversion in APON

multiple services, and OAMP functionality and scalability. As a result, a new solution has emerged—GPON (Gigabit PON), offering high bit rate support while enabling the transport of multiple services, specifically data and TDM, in native formats and with extremely high efficiency.

The gigabit-capable PON (G-PON) is specified by ITU-T G.984 series [44]:

- G.984.1—covers the requirements and basic architecture;
- G.984.2—specifies the physical-medium-dependent (PMD) layer;
- G.984.3—specifies the G-PON transmission convergence (TC) layer;
- G.984.4—standardizes the G-PON management requirements;
- G.984.5—defines enhancement band;
- G.984.6—defines reach extension;
- G.984.7—defines long reach.

Besides this, a few amendments have reached consent by the ITU-T on most of the documents in the series. GPON offers 2.488 Gbps bandwidth and direct support of both TDM (POTS and E1) and Ethernet traffic at the edge of the network with possible triple-play voice, data, and video services on the same PON as shown in Fig. 3.10. Moreover, through the GFP-based adaptation method, GPON offers a clear migration path for adding services onto the PON without disrupting the existing equipment or altering the transport layer in any way. In contrast to both APON and EPON which require a specific adaptation method for each service and the development of new methods for emerging service, the core foundation of GPON is a generic adaptation method, which already covers adaptation schemes for any possible service.

Namely, GPON is optimized for TDM traffic and relies on framing structures with a very strict timing and synchronization requirements, as it was the case with BPON, and at the same time, it offers more efficient IP and Ethernet handling.

Fig. 3.10 Native protocol
transport in GPON

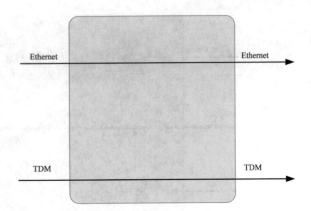

Moreover, GPON maintains the same optical distribution network and wavelength plan as BPON. In the GPON, the ITU defined a way of hiding the underlying existence of 53-byte cells by enveloping multiple cells into more reasonable large packet sizes called GPON encapsulation method (GEM) frames. The inherited 125-ms framing remains, but the cells can be optioned out. In this way, GPON achieves higher bandwidth and higher efficiency using larger, variable-length packets. GPON offers efficient packaging of user traffic, with frame segmentation allowing higher QoS for delay-sensitive voice and video communications traffic. The major characteristics of GPON include [45] (Figs. 3.11 and 3.12)

- Full service support including voice, TDM, Ethernet (10/100 BaseT), ATM, leased lines, and wireless extension;
- Physical reach of at least 20 km with a logical reach support within the protocol of 60 km;
- Support for various bit rate options using the same protocol, including symmetrical 1.25 and 2.5 Gbps and asymmetrical downstream/upstream 2500/625 and 1250/155 Mbps;
- Strong OAMP capabilities offering end-to-end service management;
- Security at the protocol level for downstream traffic due to the multicast nature of PON;
- GPON employs GEM to enable packet fragmentation. This method uses a complicated algorithm to delineate variable-size GEM segments and reconstruct the packets at the receiving device;
- High efficiency with no overhead transport of required native TDM traffic;
- Dynamic allocation of upstream bandwidth via bandwidth maps (pointers) for each ONT.

The GPON network architecture supports a two-wavelength WDM scheme for downstream and upstream digital services as shown in Fig. 3.11. Additionally, another downstream wavelength is allocated for the distribution of analog video service. As already stated, the network supports up to 60 km reach, with 20-km differential reach between optical network units (ONUs). The split ratio supported

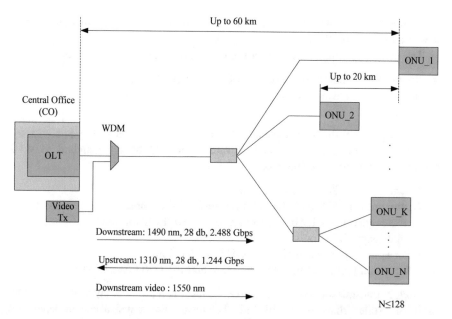

Fig. 3.11 GPON transmission

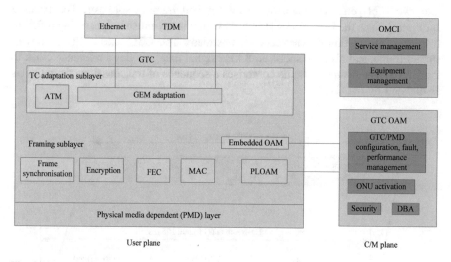

Fig. 3.12 GPON layering

by the standard is up to 128 but practical deployments would typically have lower reach and split ratios, limited by the optical budget.

ITU-T G.984.2 specifies the PMD layer for GPON, covering the range of GPON upstream and downstream bit rates, and the optical parameters for various rate combinations [44]. Today, the preferred GPON bit rate is selected to be 2.488 Gbps downstream, and 1.244 Gbps upstream. In GPON, a frame is a 125-ms quantum as it was a case in BPON as shown in Fig. 3.12.

The layer structure of the GPON OLT or ONU consists of two layers:

- the physical-medium-dependent (PMD) layer [G.984.2];
- the GPON transmission convergence (TC or GTC in the literature) layer [G.984.3].

GPON physical-medium-dependent layer, unlike the higher layers, is composed of hardware, not software. This hardware is defined by the standard G.984.2 obeying the following parameters:

- Downstream: 1.24416, 2.48832 Gbps;
- Upstream: 0.15552, 0.62208, 1.24416 or 2.48832 Gbps;
- Wavelengths: from 1260 to 1360 nm up, 1480–1500 nm down;
- Traffic type: digital only;
- Fiber splits: up to 64, limited by ODN attenuation;
- Attenuation permitted between the OLT and the ONU (the ODN).

G-PON transmission convergence (GTC) layer is a protocol layer of the G-PON protocol suite that is positioned between the physical-media-dependent (PMD) layer and G-PON clients. The GTC layering and the main functions of the user and control planes are shown in Figs. 3.12 and 3.13. The GTC is divided into two sublayers: the adaptation sublayer and framing sublayer. The framing sublayer defines the GTC frame structure, which is asymmetrical, carrying different overhead information downstream vs. upstream. The GTC uses a 125-μs downstream frame, and also transports an 8 kHz signal that provides a reference clock to the ONUs. The upstream frame comprises a sequence of transmissions from ONUs as dictated by the OLT.

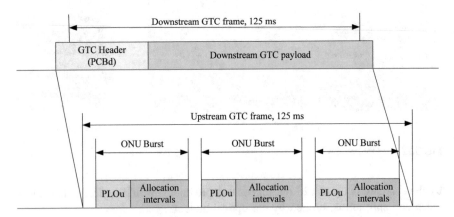

Fig. 3.13 GTC layer framing

The GTC layer provides:

- the medium access control (MAC) function;
- coordinating the interleaving of upstream transmissions from multiple ONUs;
- control functions:

  - protocol and procedures for registering ONUs to the GPON network;
  - monitoring ONUs performances;

- transport features:

  - forward error correction (FEC);
  - encryption;
  - bandwidth allocation.

The GTC layer is composed of GTC framing sublayer and GTC adaptation sublayer. This layer performs the adaptation of user data onto the PMD layer [44]. In accordance with the standard, GTC framing sublayer is a sublayer of the G-PON transmission convergence layer that supports different functions of the GTC frame as well as embedded OAM functions. Namely, embedded OAM is an operation and management channel between the OLT and the ONUs that utilizes the structured overhead fields of the downstream GTC frame and upstream GTC burst, and supports time-sensitive functions, including bandwidth allocation, key synchronization, and DBA reporting. Moreover, within the framing sublayer, the standard defines the physical layer OAM (PLOAM) function. PLOAM function is defined as a message-based operation and management channel between the OLT and the ONUs that supports the PON TC-layer management functions, including ONU activation, OMCC (ONU management and control channel) establishment, encryption configuration, key management, and alarm signaling.

GTC framing sublayer has the following three functionalities:

- Multiplexing and de-multiplexing: PLOAM and GTC payload sections are multiplexed into a downstream GTC frame per the specified frame format. In the upstream direction, each section is extracted from an upstream burst according to the defined bandwidth map corresponding to the upstream GTC frame to which the burst belongs;
- Header creation and decoding: GTC frame header is created and is formatted in a downstream frame. Upstream burst header is decoded and embedded OAM is performed;
- Internal routing function based on Alloc-ID: Routing based on Alloc-ID is performed for data from/to the GEM TC adapter.

The C/M (Control/Management) plane includes OMCI (ONT management and control interface) and GTC OAM (operations, administration, and maintenance). OMCI incorporates a full ONU management information base (MIB), and the OMCC function which conveys MIB information between the OLT and ONUs. The MIB comprises a set of managed entities, where the creation of managed entities and their attributes is designated to either the OLT or the ONU.

Figure 3.13 shows the GTC frame structure for downstream and upstream directions. The downstream GTC frame consists of the physical control block downstream (PCBd) and the GTC payload section. The upstream GTC frame contains multiple transmission bursts. Each upstream burst consists of the upstream physical layer overhead (PLOu) section and one or more bandwidth allocation interval(s) associated with a specific Alloc-ID. The downstream GTC frame provides the common time reference for the PON and the common control signaling for the upstream.

A detailed diagram of the downstream GTC frame structure is shown in Fig. 3.14. The frame has a duration of 125 µs and is 38,880 bytes long, which corresponds to the downstream data rate of 2.48832 Gbps. The PCBd length range depends on the number of allocation structures per frame. As previously stated, the downstream GTC comprises the PCBd, a header containing all overhead fields, followed by the payload part. The PCBd includes fields related to framing and physical layer operations, administration, and maintenance (PLOAM) field as shown in Fig. 3.14. The PLOAM carries a message-based protocol for PMD and GTC layer management. Finally, the PCBd includes the bandwidth (BW) map field specifying the ONUs' upstream transmission allocation. Payload may have ATM and GEM partitions (either one or both).

The upstream GTC frame is a 125-µs interval with well-defined boundaries containing multiple upstream transmission bursts controlled by the individual BW maps. In G-PON systems with the 1.24416 Gbps uplink, the upstream GTC frame size is 19,440 bytes; in G-PON systems with 2.48832 Gbps uplink, the GTC frame size is 38,880 bytes. Each upstream frame contains a number of transmission bursts coming from one or more ONUs as shown in Fig. 3.15. Each upstream transmission burst contains an upstream PLOu section and one or more bandwidth allocation intervals associated with individual Alloc-IDs. The PLOu section is sent at the beginning of any transmission burst of an ONU. The BW map field defines the arrangement of the bursts within the frame and the allocation intervals within each burst. Each allocation interval is controlled by a specific allocation structure of the BW map. A bandwidth allocation interval may contain two types of GTC layer overhead fields:

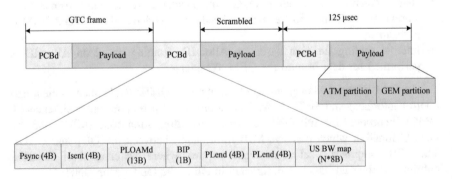

Fig. 3.14 Downstream GTC frame structure

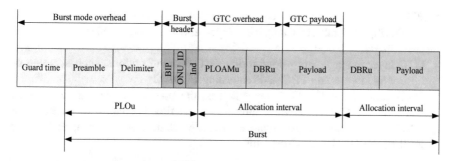

Fig. 3.15 Upstream GTC frame structure

- upstream physical layer operations, administration, and management (PLOAMu) message (default Alloc-IDs only);
- upstream dynamic bandwidth report (DBRu).

The OLT uses the BW map field to indicate to the ONU to include the PLOAMu and/or DBRu fields into the corresponding upstream allocation interval. The OLT should request PLOAMu transmission only in allocation intervals associated with the default Alloc-ID of any given ONU. The DBRu structure contains information that is tied to a T-cont entity, as opposed to the ONU as a whole. This field is sent when the corresponding flags are set in the appropriate allocation structure within the BW map.

Besides the previously described fields, the upstream frame also contains the following fields:

- BIP —bit interleaved parity field. The BIP field is an 8-bit field that contains the bit interleaved parity of all bytes transmitted since the last BIP (not including the last BIP) from this ONU, excluding the preamble and delimiter bytes, and FEC parity bytes (if present);
- ONU_ID—The ONU_ID field is an 8-bit field that contains the unique ONU-ID of the ONU that is sending this transmission. The ONU_ID is assigned to the ONU during the ranging process;
- Ind—Indication field provides real-time ONU status reports to the OLT.

The OLT bandwidth allocation method for ONU upstream transmission may be static or dynamic (DBA). Two methods of DBA are defined for G-PON:

- Status-reporting DBA, which is based on ONU reports via the DBRu field;
- Non-status-reporting DBA, which is based on OLT monitoring per T-cont utilization.

The GTC layer control plane is mainly operated via the PLOAM message protocol and some overhead fields referred to as embedded OAM. The following management functions are included:

- PMD layer management—configuration of upstream overhead as well as monitoring the health of the physical layer, and generation of alarms or statistics accordingly;
- GTC layer management—configuring GTC framing options, such as usage of upstream/downstream FEC and requesting PLOAM;
- ONU activation—The GTC layer defines the process which will activate an ONU on the GPON network, including a ranging procedure to measure the ONU distance and set its equalization delay. If necessary, the optical power level of the ONU may also be tuned.

Through the TC adaptation sublayer, the GTC layer defines two adaptation methods for data transport (Figs. 3.16 and 3.17):

- Asynchronous transfer mode (ATM);
- GPON-encapsulation method (GEM).

Accordingly, there are two flows of GPON user frames, one into and one out of a node from the PON ODN, namely the flow of frames containing ATM cells and GEM frame flows. As explained in the previous section, BPON carries only one

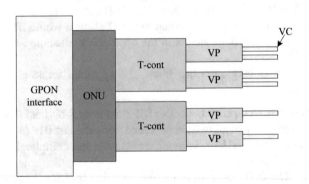

Fig. 3.16 ATM adaptation method

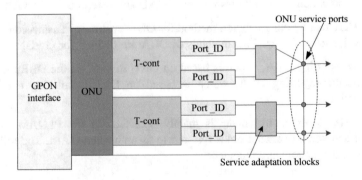

Fig. 3.17 GEM adaptation method

type of user traffic flows, i.e., cells. Today, GEM represents a preferred method that allows low overhead adaptation of various protocols, including Ethernet and TDM. In order to support these transmissions, an ONUs must somehow be labeled and identified in the network. GPON has several types of labels:

(1) ONU_ID (1B);
(2) Transmission-container (12b), where multiple T-conts can be defined per ONU:

- For ATM mode (Fig. 3.13)

    – Virtual path identification (VPI);
    – Virtual circuit identification (VCI);

- For GEM mode (Fig. 3.14)

    – Port_ID (12b) (12b).

GEM is a data frame transport scheme used in GPON systems that is connection-oriented and supports fragmentation of user data frames into variable-size transmission fragments. This is achieved through the definition of the GEM port. Namely, GEM port is an abstraction on the GTC adaptation sublayer representing a logical connection associated with a specific client packet flow. Within the GEM port definition, GTC adaptation sublayer supports GEM encapsulation, GEM frame delineation, and GEM Port_ID filtering.

GEM scheme defines a protocol-independent connection-oriented encapsulation for variable-size packets. As previously stated, GEM's virtual connection unit is called a GEM port, and may contain a flow to/from a physical or logical port of an ONU. GEM frames include a 5-byte header indicating the port ID and the length of the frame. GEM port identifier, or GEM Port_ID, is a 12-bit number that is assigned by the OLT to the individual logical connection. The GEM Port_ID assignment to the OMCC logical connection is performed by means of Configure_Port_ID PLOAM message. All other GEM Port_ID assignments for the ONU are performed via OMCC.

The format of the GEM header is shown in Fig. 3.18. The GEM header contains the payload length indicator (PLI), Port_ID, payload type indicator (PTI), and a header error control (HEC) field:

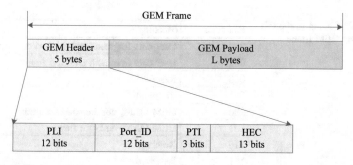

Fig. 3.18 GEM frame structure

- The PLI indicates the length L, in bytes, of the payload following this header. The PLI is used to find the next header in the stream in order to provide delineation. The 12-bit size of this field permits fragments of up to 4095 bytes. If the user data frames are larger than this, then they will have to be broken into fragments that are 4095 bytes or smaller;
- The Port_ID is used to provide 4096 unique traffic identifiers on the PON in order to provide traffic multiplexing. Each Port_ID contains a user transport flow. There can be one or more Port-IDs transmitted within an Alloc-ID/T-cont;
- The PTI field is used to indicate the content type of the payload and its appropriate treatment. The coding is shown in Table 3.2.

GEM ports are bundled onto T-conts as shown in Fig. 3.17. Transmission container is defined as a traffic-bearing object within an ONU that represents a group of logical connections. T-cont is managed via the OMCC and is treated as a single entity for the purpose of upstream bandwidth assignment on the PON. Namely, a T-cont is the unit of upstream bandwidth allocation by the OLT. The T-cont arrangement is configurable by the OLT. However, popular schemes are a single T-cont per ONU, or multiple T-conts, one per service class, per ONU [45]. As shown in Fig. 3.17, T-conts bundle GEM ports and, for example, within multiservice environment, it may be defined per CoS. One T-cont serves as bandwidth allocation unit and is identified by Alloc_ID parameter. Furthermore, GEM ports (identified by Port_ID) contain flows from logically/physical ports. The role of service adaption block is to map payload over GEM.

GEM frames may be fragmented, and hence, a client packet may span multiple GEM frames as shown in Fig. 3.19. Namely, the GEM flow consists of 125-ms-long GPON frames, whose length may or may not match that of the client user. If they do not, encapsulation is required. If the user packets are shorter than the GPON frame, it is encapsulated. If the user frame is longer than the GPON frame, it is fragmented and placed into the successive GPON frames.

GEM traffic is carried over the GTC protocol in transparent fashion as shown in Fig. 3.20. The GEM protocol has two functions: to provide delineation of user data frames and to provide port identification for multiplexing. Note that the term 'user data frames' denotes frames either going to or coming from a user.

Table 3.2 Coding in the PTI field of the GEM frame	PTI code	Meaning
	000	User data fragment, not the end of a frame
	001	User data fragment, end of a frame
	010	Reserved
	011	Reserved
	100	GEM OAM, not the end of a frame
	101	GEM OAM, end of a frame
	110	Reserved
	111	Reserved

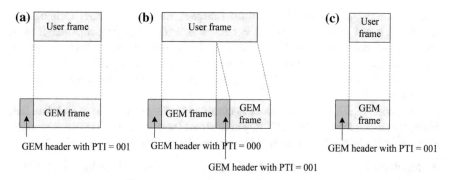

Fig. 3.19 Encapsulation of downstream user frames into GEM frames: **a** user frame matches GEM frame; **b** user frame longer than GEM frame; **c** user frame shorter than GEM frame

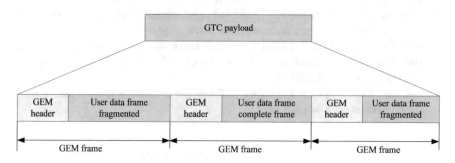

Fig. 3.20 Mapping of GEM frames into GTC payload

3.2.3 Ethernet PON (EPON)

The EPON proposal is currently being studied within the IEEE 802.3ah working group, through the task force. 'Ethernet in the First Mile' project includes three approaches [46]:

- Point-to-point copper over the existing copper plant at speeds of at least 10 Mbps up to at least 750 m;
- Point-to-point fiber over a single fiber at a speed of 1 Gbps up to at least 10 km;
- Point-to-multipoint fiber using PON at a speed of 1 Gbps up to at least 10 km.

The third approach is referred to as EPON. In the following section, we discuss in detail the characteristics and implementation of EPONs. Here, we emphasize only the main characteristics of EPONs in order to make the comparative analysis between EPONs on the one side, and BPON, and GPON on the other side.

Ethernet PONs evolved from ATM-based passive optical networks. This evolution is basically data-centric since most of data traffic within enterprises is IP-/ Ethernet-based. EPON/Gigabit EPON is governed by IEEE and is designated as

IEEE 802.3ah [46]. EPON is based on Ethernet, unlike other PON technologies which are based on ATM. Ethernet is chosen as a transport protocol since it is an inexpensive technology that is ubiquitous and interoperable with a wide variety of legacy equipment. These choices are a step forward in making PON more suitable for delivering Internet Protocol (IP)-based applications and multimedia traffic to end-users [45].

The scope of IEEE 802.3ah work is confined to the physical layer and data link layer, two lower layers of the OSI (Open Systems Interconnection) reference model. Each of these layers is further divided into sublayers and interfaces (Fig. 3.21):

- Medium-dependent interface (MDI) specifies the physical medium signals and the mechanical and electrical interface between the transmission medium and physical layer devices;
- Physical-medium-dependent (PMD) sublayer is located just above the MDI, and it is responsible for interfacing to the transmission medium;
- Physical medium attachment (PMA) sublayer contains the functions for transmission, reception, clock recovery, and phase alignment;
- PCS (Physical Coding Sublayer) is the layer that deals with line coding. The IEEE 802.3ah standard defines RS block codes in the PCS layer, and the same code is used in GPON;
- The RS (Reconciliation Sublayer) includes the emulation layer which creates virtual private path to each ONU.

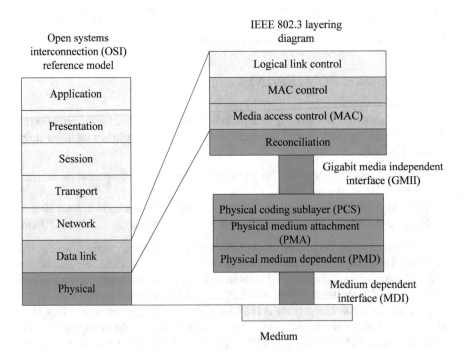

Fig. 3.21 Relationship of 802.3 layering model to OSI reference model

The data link layer consists of the following sublayers (from lower to higher):

- MAC sublayer defines a medium-independent function responsible for transferring data to and from the physical layer. The MAC sublayer defines data encapsulation such as framing, addressing, and error detection as well as medium access such as collision detection and deferral process;
- MAC control sublayer is an optional sublayer performing real-time control and manipulation of MAC sublayer operation. The MAC control structure and specification allows new functions to be added to the standard in the future;
- Logical link control (LLC) sublayer defines a medium access-independent portion of the data link layer. This sublayer is outside the scope of IEEE 802.3. Correspondingly, MAC and the optional MAC control sublayer are specified in such a way that they are unaware whether LLC, or any other client, is located above them.

According to the EPON PMD sublayer specification, EPON technology provides bidirectional 1 Gbps links using 1490-nm wavelength for downstream and 1310 nm for upstream, with 1550 nm reserved for future extensions or additional services, such as analog video broadcast. Supported fiber splits are in the range from 16 to 64, where for more than 16 splits forward error correction (FEC) functionality has to be implemented. The supported range goes from 0 to 20 km.

Furthermore, EPON is a TDM-based PON that carries all data encapsulated in Ethernet frames and is backward compatible with the existing IEEE 802.3 Ethernet standards, as well as other relevant IEEE 802 standards. Moreover, it provides simple, easy-to-manage connectivity to low-cost Ethernet-based IP equipment both at the customer premises and at the central office. Furthermore, EPON is a TDM-based PON that carries all data encapsulated in Ethernet frames and is backward compatible with the existing IEEE 802.3 Ethernet standards, as well as other relevant IEEE 802 standards (Fig. 3.22).

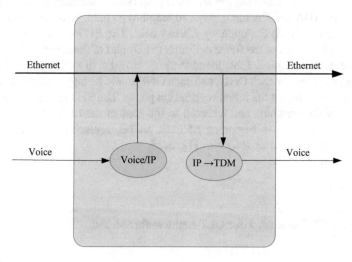

Fig. 3.22 Protocol conversion in EPON

The IEEE 802.3ah task force has standardized MPCP (Multipoint Control Protocol) protocol to control the communication and data exchange in EPON [46]. EPON is based upon a mechanism named MPCP, defined as a function within the MAC control sublayer. MPCP uses messages, state machines, and timers, to control access to a point-to-multipoint (P2MP) topology. Each ONU in the P2MP topology contains an instance of the MPCP protocol, which communicates with an instance of MPCP in the OLT. The introduction of the MPCP was necessary for the standardization of EPON under the IEEE 802.3 Ethernet standard and for ensuring compliance with all of the requirements put forward by the standard. Namely, within the standard, all Ethernet stations interconnected by a shared medium should form an access domain and be able to communicate with each another. On the other hand, EPON was being developed for subscribers in access networks with requirements drastically different from those of private LANs, i.e., subscribers in the access networks are independent users which cannot communicate to each another, except when provisioned by a service provider. To resolve this issue and to ensure seamless integration with other Ethernet networks, devices attached to the PON medium implement a logical topology emulation (LTE) function. The LTE function, based on its configuration, may emulate either a shared medium or a point-to-point medium which will be described in detail in the next section.

To preserve the existing Ethernet MAC operation defined in the IEEE 802.3 standard, the LTE function should reside below the MAC sublayer. The basis of the EPON/MPCP protocol lies in the point-to-point (P2P) emulation sublayer, which makes an underlying P2MP network appear as a group of P2P links from the viewpoint of the higher protocol layers (at and above the MAC Client). EPON does not use encapsulating framing in either the upstream or downstream direction; instead, the content of the Ethernet preamble is modified. Namely, standard Ethernet starts with an essentially content-free 8B preamble where EPON overwrites some of the preamble bytes in order to hide the new PON header. The generic Ethernet frame and the EPON frame both contain the destination and source MAC address (DA and SA filed), payload length/type field, a variable-size payload field, and CRC (Cyclic Redundancy Check) field. The EPON header modifies the preamble and the start of the frame delimiter (SFD) part of the Ethernet MAC frame to include LLID (Logical Link Identification) in order to enable P2PE function. LLID tag is unique for each ONU, and each ONU could be tagged with one or more tags by the OLT during the initial registration phase. The SFD field is moved to the third byte of the preamble and renamed to the start of the LLID delimiter (SLD). LLID is prepend to the beginning of each packet, replacing two octets of the Ethernet preamble field as shown in Fig. 3.23.

LLID field contains:

- MODE (1b);

  - always 0 for ONU;
  - 0 for OLT unicast, 1 for OLT multicast/broadcast;

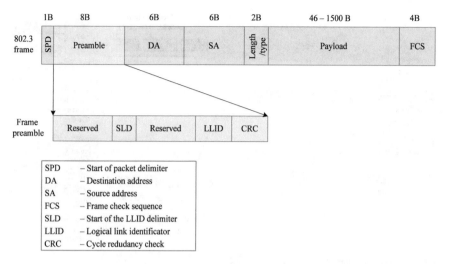

Fig. 3.23 Format of the frame preamble in EPON

- actual logical link ID (15b);
  - identifies registered ONUs;
  - 7FFF for broadcast.

In downstream transmission, the LLID field defines the destination ONU. An ONU filters the received frames based on the LLID in the frame's preamble and its own unique LLID value assigned by the OLT as shown in Fig. 3.24. A special value is reserved for broadcast messages sent to all ONUs. In upstream transmissions, the LLID field marks the source ONU (Fig. 3.25), where CRC field is used to validate preamble integrity. In practice, most ONU equipment is registered by the OLT as a single ONU which uses a single LLID for data transport [47]. However, some equipment is registered as multiple virtual ONUs, thereby establishing multiple LLIDs. This allows EPON to access the same traffic granularity on the PON as GPON.

The communication and data exchange in EPON is defined as follows. In the downstream direction, an EPON operates as a broadcast network where OLT broadcasts frames to all ONUs and only the ONU that has the right MAC address and proper LLID of the frames will receive them. The actual connection between ONU MAC address and the defined LLID identification depends on the implemented LTE mode (shared medium or a point-to-point medium) and will be described in the next section. On the other hand, in the upstream direction, a TDMA-based PON is a multipoint-to-point network, where multiple ONUs share the same transmission channel. As a result, data streams transmitted simultaneously from different ONUs may still collide. Hence, access to the shared medium must be arbitrated by MAC protocol to prevent collisions between Ethernet frames of different ONUs transmitting simultaneously.

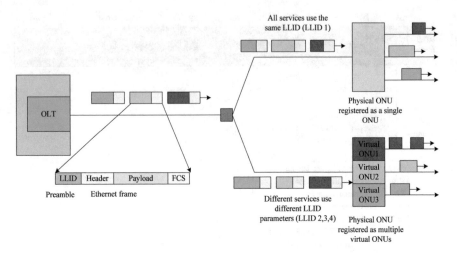

Fig. 3.24 Downstream transmission in EPON

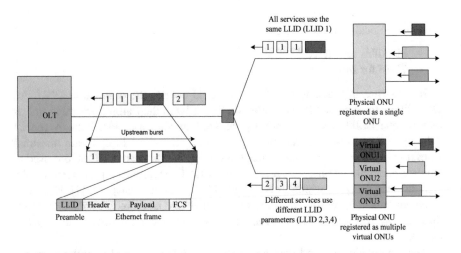

Fig. 3.25 Upstream transmission in EPON

As already stated, the IEEE 802.3ah task force has standardized MPCP protocol to control data exchange in EPON. MPCP allows for the sharing of upstream bandwidth among various ONUs and prevents collisions of packets originating from different ONUs. MPCP is based upon TDM and allows OLT to arbitrate between various ONUs requesting upstream transmission over the shared medium by assigning exclusive timeslots to individual ONUs. Each ONU can transmit packets upstream only during its assigned timeslot(s). MPCP is not concerned with any particular bandwidth allocation; it is merely a supporting protocol that facilitates the implementation of various bandwidth allocation algorithms in EPON [48].

Besides the simplified communication based on the transmissions of native Ethernet frames, OAM functionality is another important EPON breakthrough [47]. Ethernet now includes link layer management that enables OLT to remotely manage attached ONUs. OAM is established after the discovery process and is maintained by periodic message transmission. For example, information about remote failures is conveyed using flags in OAM messages to indicate failure status or the remote ONU can be instructed to return incoming packets as part of the remote loopback functionality. Moreover, OAM link information can be extended beyond the OLT by placing a SNMP (Simple Network Management Protocol) agent at the OLT. In addition to the improvement in management functionality, today, all EPON implementations incorporate AES encryption [43].

3.2.4 Comparison of BPON, GPON, and EPON Technologies

The main characteristics and comparison of the BPON, GPON, and EPON technologies are shown in Table 3.3.

The shortcomings of BPON technology have already been analyzed in Sect. 3.2.1. As described in the previous sections, the transmission of IP traffic over an ATM system requires segmentation into 53-byte cells (five cells out of which are overhead) which is a time-consuming and inefficient process. Carrying IP traffic over Ethernet over ATM requires a complicated adaptation layer (implementing

Table 3.3 BPON, GPON, and EPON technology comparison

Parameter	BPON (G.983)	GPON (G.984)	EPON (802.3ah)
Bit rates downstream	155 and 622 Mbps	1.2 or 2.4 Gbps	1.25 Gbps
Bit rates upstream	155 Mbps	155, 622 Mbps, 1.2 or 2.4 Gbps	1.25 Gbps
Packet capability	Cells only	Fragmentation every 125 μs	Native
Analog video	Specified	Not specified	Not specified
Protection switching	Specified	Not specified	Not specified
Encryption	Not specified	AES	Not specified
Number of ONUs	Up to 32	Up to 128	Up to 64
Error protection	Control fields, polynomial code for each cell	Control fields, no line code	Preamble, line code, FEC
Address space	8 bits	8 bits	48 bits
Class of service	5 T-cont types	5 T-cont types	8 queues

RFC 1483), adding more cost and complexity to the OLT and ONTs. On the other hand, considering that the variable-length structure of the Ethernet frame is up to 1518 bytes for the transfer of IP traffic, Ethernet represents an ideal solution for transporting IP data, and Ethernet devices are optimized for data networks. In addition, EPON is based on the transmission of native Ethernet frames and it is able to efficiently support the implementation of different multilayered security mechanisms for IP traffic, such as firewalls, VPNs, Internet Protocol security (IPSec), and tunneling [45]. Apart from the efficiency issues, there is also a large price difference between Ethernet systems and ATM-based systems.

With regard to GPON and EPON differences, the most dramatic distinction between the two protocols is a marked difference in the architectural approach [49]. Namely, a GPON consists of multiple Layer 2 networks over the same physical layer where each network has a different L2 transmission protocol. As shown in Fig. 3.26, with multiprotocol transport solution, GPON provides three Layer 2 networks: ATM for voice, Ethernet for data, and proprietary encapsulation for voice. EPON, on the other hand, employs a single Layer 2 network that uses IP to carry data, voice, and video. The support for ATM technology allows provisioning of virtual circuits for different types of services sent from a central office location primarily to business end-users. Although this type of transport provides high-quality service, it involves significant overhead because virtual circuits need to be provisioned for the each type of service. Additionally, in the case of ATM support, GPON has to support, among others, multiple protocol conversions, segmentation and reassembly (SAR), virtual channel (VC) termination, and point-to-point protocol (PPP).

On the other hand, EPON provides seamless connectivity for any types of IP-based or other 'packetized communications' (Fig. 3.27). Moreover, Ethernet devices are widely used from the home network, so the implementation of EPONs is highly cost-effective. The use of EPON allows carriers to eliminate complex and expensive ATM and SONET elements and to simplify their networks, thereby lowering costs to subscribers. Furthermore, an EPON equipment cost is lower in comparison with the GPON equipment and is hence widely used in Ethernet networks of today.

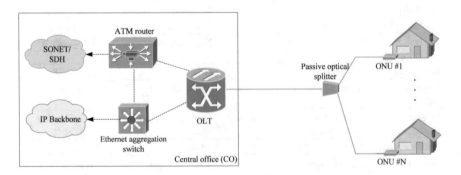

Fig. 3.26 Multiprotocol transport support in GPON

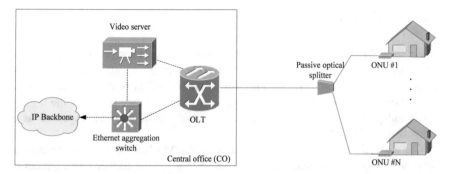

Fig. 3.27 Transmissions in the EPON network

Since EPON requires a single management system, versus three management systems for the three Layer 2 protocols in GPON, this means that EPON results in a significantly lower total cost of ownership [49]. Moreover, since EPON does not require multiprotocol conversions, this results in the lower network cost as well.

Within both PON protocols, a fixed overhead is added in order to transmit user data in the form of packets. As explained in previous sections, in EPONs, data transmission occurs in variable-length packets of up to 1518 bytes according to the IEEE 802.3 protocol for Ethernet. In ATM-based PONs, including GPONs, data transmission occurs in fixed-length 53-byte cells (with 48-byte payload and 5-byte overhead) as specified by the ATM protocol. Accordingly, in the ATM-based PON, one IP packet must be converted in the form of ATM cell (48-byte segments with a 5-byte header) which makes this network very inefficient for the transport of IP traffic. This process is time-consuming and complicated and adds cost to the OLT as well as to the ONUs. Hence, in today's GPONs, GEM encapsulation mode is preferred over ATM encapsulation mode. Furthermore, GPON does not support multicast services, which makes the support for IP video more bandwidth consuming. On the contrary, since EPON transmits IP packets encapsulated in the native Ethernet frames, this means that multicast support is not the issue.

With regard to security issues, encryption, within GPON, is part of the ITU standard but only for the downstream transmission. EPON, on the other hand, uses an AES-based mechanism, which is supported by multiple vendors and deployed in the field. Moreover, EPON encryption is both downstream and upstream [47].

In recent years, optical access networks based on EPON have gained a lot of attention in the industry as well as within the framework of academic research. Industrial interests have arisen from the fact that EPON represents the first optical technology that promises to be enough cost-effective to justify its mass deployment in the access network. We strongly believe that EPON is the optimal solution for the realization of optical access networks. EPON implementation, DBA algorithms as well as QoS support, will be analyzed in greater detail in the following chapters.

3.3   WDM PON

Apart from the TDMA approach, another possible way for separating the ONUs' upstream channels is to use wavelength division multiple access (WDMA). WDM PONs (Wavelength Division Multiplexing Passive Optical Networks) are the next generation in the development of access networks since they can offer the highest bandwidth and the successful transmission of multimedia real-time applications in terms of QoS and SLAs.

Currently, there is no common standard for WDM PON, but the development of these networks is considered to be the key factor for the further development of access networks. Until now, various approaches and architectures have been proposed for the realization of different flavors of WDM PONs (WDM EPON and NGN GPON among others). In one of the suggested approaches, the multiple wavelengths of a WDM PON can be used to separate ONUs into several virtual PONs that coexist on the same physical infrastructure. Alternatively, the wavelengths can be used collectively through statistical multiplexing to provide efficient wavelength utilization and lower delays experienced by the ONUs. In the other approach, WDM is defined as a PON network that supports transmission on more than one wavelength in any direction.

The most widely accepted definition of WDM PON is a definition in which each ONU operates on a different wavelength. In this approach, the architecture of WDM PON is similar to the architecture of the PON. The main difference is that multiple wavelengths operate on single fiber and ONUs operate on different wavelengths. Multiple wavelengths on single fiber enable either more bandwidth per each ONU or more ONUs per each distribution fiber (Fig. 3.28). The MAC layer is simplified because point-to-point connections between OLT and ONUs are

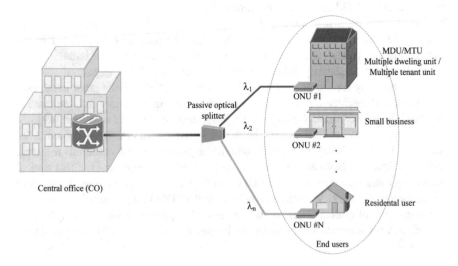

Fig. 3.28 WDM PON components

realized in the wavelength domain, so no point-to-multipoint media access control is needed. In WDM PON, each wavelength can run at a different speed and protocol, and therefore, the network can be upgraded on per-need basis. Moreover, WDM PON has better privacy and better scalability because each ONU only receives its own wavelength. The drawback of this solution is the high cost of the field implementation, since in most cases, tunable laser or multiple lasers in OLT have to be implemented.

In any case, all kinds of PON provide higher bandwidth than traditional coper-based access networks and some of them will be discussed in more detail in the following chapters.

Chapter 4
Single-Channel EPON

4.1 Introduction

As defined in the IEEE 802.3 standard, EPON is a PON-based network that carries data traffic encapsulated in Ethernet frames. According to the IEEE 802.3 standard [50], Ethernet MAC can operate in one of the two modes: CSMA/CD (Carrier-Sense Multiple Access with Collision Detection) mode or full-duplex mode. The standard defines two basic modes of operation for an Ethernet network:

- Shared-medium network where CSMA/CD protocol has to be deployed in order to avoid collisions and
- full-duplex point-to-point links where stations are connected through a Layer 2 switch.

However, EPON cannot be considered as either a shared medium or a point-to-point network since it has the characteristics of both models. As we explained in the Sect. 3.2.3, the upstream and downstream transmissions are separated.

4.2 Characteristics of the MPCP Protocol

As previously explained, in order to be part of the IEEE 802.3 standard, EPON must incorporate transmission arbitration within the MAC control sublayer, i.e., transmission based on Ethernet MAC, either CSMA/CD or full-duplex. The MAC control sublayer resides between the MAC and LLC sublayers, and initially, flow control was its only function, as shown Fig. 3.21. The implementation of the flow control operation allowed a station to pause transmission from its peer for a pre-determined interval of time with the use of a PAUSE MAC control message. However, transmission in EPON is defined as being exactly the opposite from flow

© Academic Mind and Springer International Publishing AG 2017
M. Radivojević and P. Matavulj, *The Emerging WDM EPON*,
DOI 10.1007/978-3-319-54224-9_4

control, i.e., as operation that allows a station to enable transmission from its peer for a predetermined interval of time (timeslot or transmission window).

Consequently, the IEEE 802.3ah task force [46] developed the MPCP. MPCP defines a message-based mechanism to facilitate real-time information exchange between the OLT and each ONU. Instead of specifying a particular scheduling approach, i.e., the bandwidth allocation algorithm, MPCP only provides a basic mechanism for developing a wide range of bandwidth allocation schemes. The exact choice of such schemes is left to vendors.

The implementation of MPCP mechanism in EPONs allows, among other, the following:

- Optimization of network resources;
- Support for the ranging procedure in order to determine ONU distance and decrease latency;
- Fast scheduling cycles allow support of oversubscription;
- Support for the increase in number of ONUs, i.e., 64 ONUs;
- Negotiation of parameters in order to achieve optimal performance;
- Long reach (20 km);
- Granting cycle allows low end-to-end delays and support voice services; and
- Dynamic granting capability allows fast bandwidth assignment;
- Single copy broadcast capability in downstream direction allows broadcasting of video application without bandwidth waste.

Consequently, the main functions of the MPCP include the following:

- Provision of timing reference for the synchronization of ONUs;
- Control of the autodiscovery process; and
- Bandwidth/timeslot assignment to ONUs.

In order to support the above-stated functions, the MPCP specifies the control mechanism used between an OLT and ONUs connected to a point-to-multipoint EPON segment in order to allow an efficient transmission of data in the upstream direction. MPCP uses MAC control messages (similar to the Ethernet PAUSE message) to coordinate multipoint-to-point upstream traffic.

With the implementation of the MPCP, collisions are avoided because the OLT allows only one ONU to transmit at any given time with the generation of MPCP control messages. When ONU receives the control message, the transmission is enabled for the predetermined period. Accordingly, the standard defines two MPCP modes of operation: autodiscovery (initialization) and the normal mode of operation. The normal mode, or bandwidth assigned mode, is used to assign transmission windows to all discovered ONUs. In order to sustain communication between the OLT and ONUs, in this operation mode the MPCP must provide the periodic granting for each ONU. On the other hand, the autodiscovery mode is used to detect newly connected ONUs and learn their parameters, such as MAC addresses and round-trip delays. To discover newly activated ONUs, the MPCP should initiate the discovery procedure periodically. While the MAC control sublayer is optional for other configurations, in EPON it is mandatory, because EPON cannot operate without MPCP. Hence, the definition of the protocol includes the definition of the control messages used in the autodiscovery and normal mode of operation as well as the RTT (round-trip time) measurement.

Accordingly, MPCP introduces the five new MAC control messages:

- Normal mode: GATE, REPORT for assigning and requesting bandwidth and
- Autodiscovery process: REGISTER_REQ, REGISTER, and REGISTER_ACK.

4.2.1 MPCP Control Frames

The MPCP control frames are commonly referred to in the literature as MPCP data units (MPCPDUs). Basically, all MPCP control messages are 64-byte MAC control frames consisting of the following fields (Fig. 4.1):

- Destination address (DA): The destination address field contains the 48-bit address of the station for which the frame is intended. All MPCPDUs use a globally assigned 48-bit multicast address 01-80-C2-00-00-0116, except for the REGISTER message which uses the individual MAC address of the destination ONU;
- Source address (SA): The source address field of a MAC control frame contains the 48-bit individual address of the station sending the frame. The frames originated at the OLT should use the source address associated with the MAC instance which transmitted the frame;
- Length/type: The length/type field contains the hexadecimal value 88-08. This value has been universally assigned to identify MAC control frames;

Fig. 4.1 Generic MPCP
frame format

Fields (octets)

| Destination address (6 octets or 6B) |
| Source address (6 octets) |
| Length/type = 88-08(16) (2 octets) |
| Opcode (2 bytes) |
| Timestamp (4 octets) |
| Opcode-specific field/pad (40 octets) |
| Frame check sequence (FCS) (4 octets) |

Table 4.1 Definition of the opcode field

Opcode value	Control message
00-0116	Pause
00-0216	GATE
00-0316	REPORT
00-0416	REGISTER REQ
00-0516	REGISTER
00-0616	REGISTER ACK

- Opcode: The opcode field identifies the specific MAC control frame, i.e., differentiates message types as described in Table 4.1;
- Timestamp: The timestamp field carries the value of MPCP clock corresponding to the transmission of the first byte of the destination address. The timestamp values are used to synchronize MPCP clocks in the OLT and ONUs;
- Opcode-specific fields: These fields carry information pertinent to specific MPCP functions. The portion of the payload not used by the opcode-specific fields should be padded with zeros;
- Frame check sequence (FCS): The FCS field carries a CRC-32 value used by the MAC to verify the integrity of the received frames.

The REPORT and GATE control messages are used during the bandwidth assignment operation mode and are defined as follows (Figs. 4.2 and 4.3). REPORT messages are used by ONUs to report the local queue status to the OLT. The format of the REPORT frame is shown in Fig. 4.2. The Queue #n report field transmits information about queue lengths where devices support up to eight queues, according to the standard. The reported queue lengths should be adjusted to account for the necessary frame preamble, inter-frame spacing, and FEC parity overhead, even though this additional data may not be physically present in the queue.

The GATE control frame has two functions:

- The discovery GATE message is used to advertise a discovery slot for the new, uninitialized ONUs and
- The normal GATE message is used to grant transmission to the already discovered ONU.

These two functions are defined through different opcodes. Figure 4.3 presents the formats of discovery GATE and normal GATE messages. The number of grants field can contain up to four grants. A GATE message with 0 grants does not assign a transmission window to an ONU and is only used as a keep-alive mechanism. Each grant or transmission window is represented by a pair {start time, length}.

The number of grants/flag field indicates the exact number of grants in the given GATE message, as well as some additional information, as shown Table 4.2. The three least significant bits (0–2) define the number of grants transmitted in the GATE message, and the valid values are 0 through 4.

Fig. 4.2 REPORT frame
format

Fields (octets)

Destination address (6 octets)
Source address (6 octets)
Length/type = 88-08(16) (2 octets)
Opcode (2 bytes)
Timestamp (4 octets)
Report bitmap (1 octet)
Queue #0 (2 octet)
Queue #1 (2 octet)
Queue #2 (2 octet)
⋮
Queue #8 (2 octet)
Pad (0-39 octets)
Frame check sequence (FCS) (4 octets)

The third bit, or the discovery/normal GATE bit, indicates the purpose and payload format of the message. When set to 1, this field indicates that the frame is the discovery GATE; otherwise, it is a normal GATE. The force report subfield defines a bitmap that indicates whether the OLT requests the ONU to transmit the REPORT message in any of the assigned grants.

The discovery GATE message always carries a single grant. This message should have the discovery/normal GATE bit set to 1 and should have none of the force report bits set. Therefore, in discovery GATEs, the number of grants/flag field always has the value of 00001001 binary value.

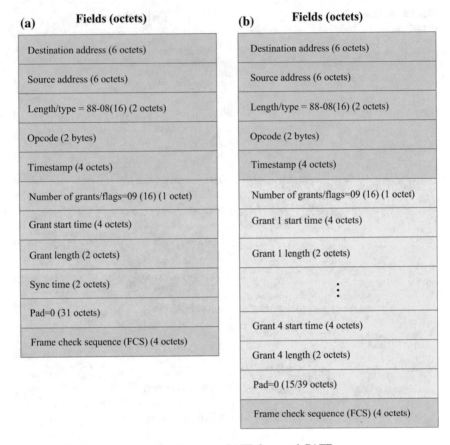

Fig. 4.3 GATE message format: **a** Discovery GATE; **b** normal GATE

Table 4.2 Contents of number of grants/flag field	Bits	Subfield name
	0–2	Number of grants
	3	Discovery/normal GATE
	4–7	Force report bitmap

4.2.2 Autodiscovery Mode

In the autodiscovery mode, the MPCP initiates the discovery procedure periodically in order to discover and initialize the newly activated ONUs. However, the autodiscovery mechanism is used not only to detect the newly connected ONUs, but also to learn the round-trip delays and MAC addresses of these ONUs and to assign LLID parameters for ONUs. Hence, the definition of the MAC control sublayer in EPON is mandatory, because EPON cannot operate without MPCP.

The implementation of the autodiscovery procedure is also mandatory since an ONU cannot turn its laser on or transmit any data unless it is granted by the OLT. Therefore, after boot up, an ONU would wait for a grant message from the OLT in order to activate. However, OLT does not know and cannot know that a new ONU exists in the system, and consequently, a new ONU cannot be integrated in the network. In resolving this situation, the autodiscovery mechanism is used to detect the newly connected ONUs and to learn the round-trip delays and their MAC addresses. The discovery process is implemented in both the OLT and ONUs and employs four MPCP control messages carried in MAC control frames: GATE, REGISTER_ REQ, REGISTER, and REGISTER_ACK.

According to the standard [46], the autodiscovery process in the OLT is defined through the definition of the four separate processes: discovery gate generation process, request reception process, register generation process, and final registration process. The definition of the separate processes allows for a definition of the one MPCP instance associated with the broadcast logical port (the OLT has to discover all newly added ONUs) and one separate instance associated with unicast logical ports (the communication between the OLT and the one defined and discovered ONU is established). On the contrary, in order to be more cost-efficient, ONUs typically have only one instance of MPCP which responds to both broadcast and unicast LLIDs.

The autodiscovery handshake procedure consists of the following steps (Fig. 4.4):

- Using the gate generation process, the OLT sends a discovery GATE message to a group of MAC control addresses through a broadcast channel. The discovery GATE MPCPDU is received and verified by the gate reception process at the ONU and stored for future activation;
- Upon initialization, the discovery process in the ONU generates a REGISTER_REQ message that remains buffered until the transmission window opens. Afterward, the REGISTER_REQ is transmitted upstream to the OLT on the broadcast channel. At the OLT, the REGISTER_REQ passes through the request reception process;
- Upon processing the REGISTER_REQ message from the ONU, the discovery agent issues a unique LLID value and requests the register generation process to transmit a REGISTER MPCPDU to the ONU. This message is addressed to an individual ONU but is transmitted on the broadcast channel, because the unique LLID has not been assigned to the ONU yet. At the ONU, the REGISTER message is forwarded to the discovery process;
- Following the transmission of REGISTER MPCPDU, the DBA agent allocates a normal grant to the newly registered ONU. This grant is needed to give the ONU an opportunity to transmit the acknowledgment back to the OLT. The final registration process issues a normal GATE MPCPDU.

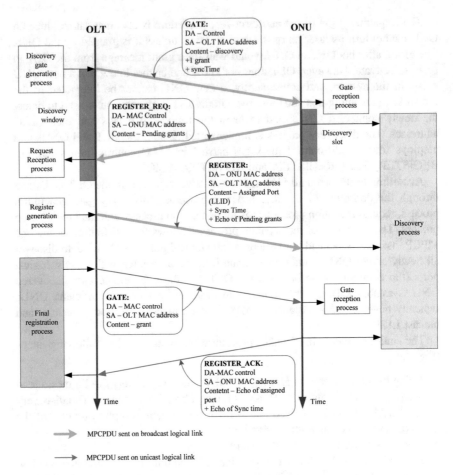

Fig. 4.4 Autodiscovery process

The discovery process in the ONU is responsible for generating REGISTER_REQ MPCPDUs, processing the received REGISTER MPCPDUs, and issuing acknowledgments in the form of REGISTER_ACK MPCPDUs.

4.2.3 Bandwidth Assignment Mode (Normal Mode)

In the normal mode, we differentiate between the upstream and downstream functions. Hence, the upstream communication uses REPORT messages, while GATE messages are used in the downstream direction, as shown in Fig. 4.5. Consequently, in the normal mode of operation, the following functions are used in the protocol layer:

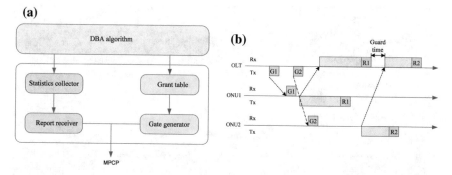

Fig. 4.5 OLT—ONU communication: **a** management block at the OLT; **b** REPORT/GATE exchange

- Bandwidth assignment—defined through the downstream communication from the OLT to the ONUs;
- Bandwidth request—defined through the upstream communication from ONUs to the OLT. At the same time, the service layer above makes the following decisions:

  - bandwidth allocation (OLT) and
  - bandwidth utilization(ONU).

As we have previously explained, the MPCP allows for the sharing of the upstream bandwidth among the various ONUs and prevents collisions of packets originating from different ONUs. Since MPCP is based upon time-division multiplexing, the OLT has to arbitrate between various ONUs requesting upstream transmission over the shared medium by assigning exclusive timeslots to individual ONUs. Therefore, each ONU can transmit packets upstream only during its assigned timeslot(s). The transmission of the two different ONUs is separated with the guard time interval, as shown in Fig. 4.5b. In the bandwidth assignment mode, the MPCP uses GATE and REPORT messages.

The GATE message, sent by an OLT to an ONU, assigns a transmission timeslot window to the ONU. The GATE message specifies the transmission start and end time during which the ONU can transmit the queued customer traffic upstream to the OLT. ONU MAC control enables PHY transmission at the start of the GATE duration and disables it at the end of the GATE duration. As previously explained, GATE messages can be used with timestamps only, as shown in Fig. 4.6. Furthermore, the intra-ONU scheduler schedules the packet transmission for various traffic queues from local users, and the transmission window may comprise multiple Ethernet frames. According to the format of control messages, the REPORT message can support eight queue reports (an ONU can support up to 8 priority queues) and the GATE message can grant four timeslots at most, as shown in Fig. 4.7.

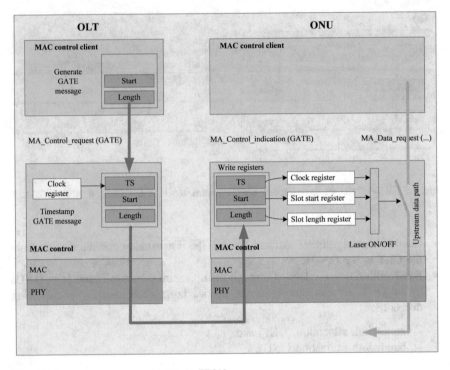

Fig. 4.6 Downstream communication in EPON

Within the allocated window, the ONU sends REPORT message along with data to report the bandwidth requirements for the upstream transmission of its traffic. REPORTs are generated in the ONU MAC control client and indicate the local conditions to the OLT. These messages are optional, but if they do exist, the OLT must process them. They can be used only with timestamps and must be issued periodically. A REPORT message is transmitted either at the beginning of or at the end of the timeslot with data packets. Based on the REPORT message that contains queue occupancy information, the DBA module (inter-ONU scheduler) in the OLT allocates the appropriate bandwidth to the ONU.

The presented architecture can easily support priority scheduling, which is a useful method for providing differentiated services which will be further discussed in the following sections.

4.2.4 RTT Measurement and MPCP Timing Model

The OLT controls the transmissions from each ONU by sending grants or permissions to them. In order to avoid collisions between transmissions from different ONUs, a technique called ranging is executed to measure the logical distance

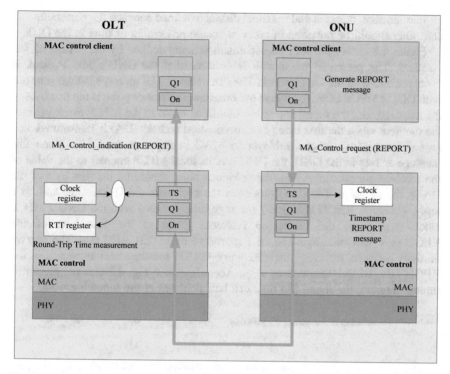

Fig. 4.7 Upstream communication in EPON

between the ONUs and the OLT. As a result, each ONU adjusts its transmission time properly, and thus, the effects of propagation delays are successfully avoided. In order to synchronize ONU's MPCP clock with the OLT's clock, each MPCP message defines a timestamp field, as shown in Figs. 4.1, 4.2, and 4.3. This field is added when message is transmitted by MAC control layer. With the timestamp field, each ONU synchronizes its local clock with the OLT clock. The OLT's control multiplexer writes the value of the MPCP clock into the timestamp field of an outgoing GATE message. When a GATE message arrives to an ONU, the local MPCP clock is set to the value received in the timestamp field.

The round-trip time compensation is another issue that has to be resolved. RTT measurement allows delay compensation in the OLT through the configuration where the grants to ONU reflect arrival time that is compensated by RTT. For example, if OLT has to receive data from an ONU at time T_o, it will send a GATE message containing:

$$\text{Start Time\_of\_the\_slot} = T_0 - RTT. \tag{4.1}$$

Furthermore, minimal and maximal distance defined between the timestamp and start time should be provided in order to enable processing of time in the OLT.

Figure 4.8 shows the RTT measurement scheme defined by the IEEE 802.3ah standard. The presented calculation is only valid if the ONU's MPCP clock is synchronized with the OLT's clock. The discovery GATE message is time-stamped with OLT's MPCP clock (t_0) where the timestamp reference point is the first byte of the message, i.e., the timestamp value should be equal to the MPCP clock value at the moment when the first byte of the destination address (DA) is transmitted, i.e., passed from MAC control sublayer to MAC sublayer at the OLT. When this message arrives at the ONU, the ONU sets its local MPCP counter to the value of the received timestamp. Again, the reference point should be the first byte of the DA in the message. The ONU processes the discovery GATE message and, as a reply, generates REGISTER_REQ message time-stamped with the ONU's MPCP clock (t_1). Now, the timestamp reference point is the first byte of the REGISTER_REQ message. The OLT receives REGISTER_REQ at t_2. The interval of time between receiving the discovery GATE message and transmitting the REGISTER_REQ is denoted as T_{wait}. According to e Fig. 4.8 and the presented timing diagram, the round-trip time can be calculated in the following way:

$$RTT = T_{downstream} + T_{upstream} = T_{response} - T_{wait} = (t_2 - t_0) - (t_1 - t_0) = t_2 - t_1.$$
$$(4.2)$$

The RTT equals the difference between the REGISTER_REQ arrival time and the timestamp contained in the REGISTER_REQ message. The presented calculation can be used only when MPCP clocks of the OLT and ONUs are synchronized [42].

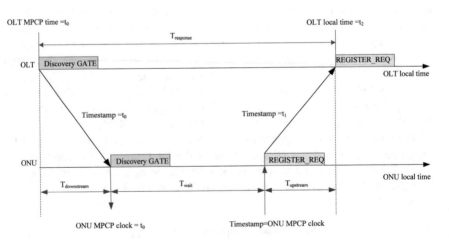

Fig. 4.8 RTT measurement

4.2.5   *Logical Topology Emulation*

As referred to in Sect. 3.2.3, in order to preserve the existing Ethernet MAC operation defined in the IEEE 802.3 standard, the standard defines the LTE function that should reside below the MAC sublayer. LTE function tags the Ethernet frames with unique logical link identifiers (LLIDs) for each ONU. In order to guarantee the unique LLID, during the initial registration (autodiscovery) phase, the OLT assigns one or more tags to each ONU. Furthermore, the LTE function, based on its configuration, may emulate either a shared medium or a point-to-point medium.

A.   *Point-to-point emulation (P2PE)*

The primary goal of P2PE mode is to achieve the same physical connectivity which is similar to switched LAN, where all the stations are connected to a central switch using point-to-point links, as shown in Fig. 4.9. In this mode, the OLT must have one MAC protocol interface for each ONU. During ONUs registration, a unique LLID value is assigned to each ONU. Moreover, each MAC port at the OLT must have the same LLID as its corresponding ONU.

In the case of downstream transmission, the emulation function in the OLT inserts the LLIDassociated with a particular MAC port from which the frame initially arrived, as shown in Fig. 4.9a. Even though the frame will pass through a splitter and reach each ONU, only one P2PE ONU function will match that frame's LLID with the value assigned to the ONU. That ONU accepts the frame and forwards it to its MAC sublayer for further verification. LTE functions in all other ONUs will discard this frame, and the MAC sublayers will never see that frame.

On the other hand, in the case of upstream transmission, each ONU inserts the LLID previously assigned by the OLT and sends the frame to the OLT. The OLT,

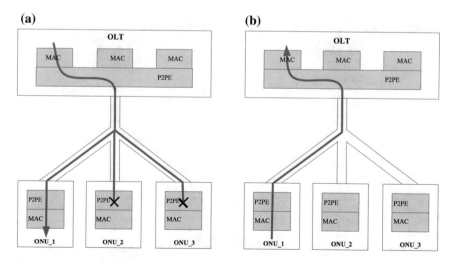

Fig. 4.9 Point-to-point communication: **a** downstream transmission; **b** upstream transmission

i.e., the configured P2PE function, demultiplexes the received frame to a proper destination port based on the received LLID value, as shown in Fig. 4.9b. It is obvious that P2PE function emulates the classic bridging function in which each ONU is virtually connected to an independent bridge port placed in the OLT. The P2PE function then relays inter-ONU traffic between the defined logical ports.

B. *Shared-medium emulation (SME)*

In the SME mode, the filtering applied in ONUs is opposite to that in the P2P mode. In the P2PE, an ONU only accepts a frame with LLID which matches ONU's own LLID, while in the SME mode, ONUs accept frames with LLIDs which are different from the ONU's assigned LLID. This means that in upstream direction, frames transmitted by any ONU should be received by every stations (OLT and every other ONU), except the sender. In the downstream direction, the OLT inserts a broadcast LLID, which will be accepted by every ONU, as shown in Fig. 4.10.

The SME requires only one MAC port in the OLT and presents a single access domain; i.e., data frame sent by any ONU will reach every other ONU. This behavior is not desirable in the access network for many reasons, and today, this concept is entirely abandoned. Moreover, the whole concept that defined the implementation of two emulation mode was not successfully accepted.

4.2.6   Final MPCP Deployment

In order to achieve optimal operation, the IEEE 802.3ah task force has considered the use of both P2PE and SME modes simultaneously, as shown in Fig. 4.11. Since both of the presented models have drawbacks, these drawbacks must be carefully

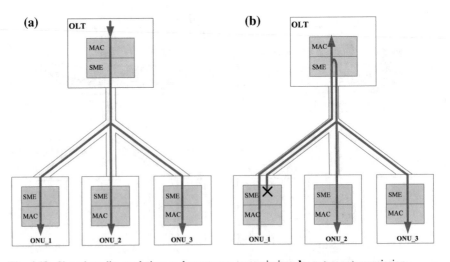

Fig. 4.10 Shared media emulation: **a** downstream transmission; **b** upstream transmission

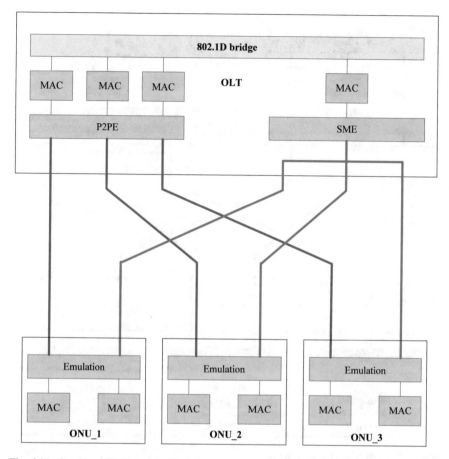

Fig. 4.11 Combined P2PE and SME mode

considered in order to be able to support various applications and multimedia services in the access network.

The P2PE mode does not support native broadcast, and hence, the OLT in the P2PE mode must duplicate the broadcast packets each time with a different LLID. Having in mind that multimedia services have become more and more present in the access network environment, broadcast/multicast support has consequently become the key factor for the realization of the multiservice environment. On the other hand, the SME provides broadcast capabilities but also reflects every upstream frame in the downstream direction. Consequently, a large portion of downstream bandwidth is wasted. To achieve optimal operation, the IEEE 802.3ah task force has considered the possibility of using both P2PE and SME modes simultaneously. However, the SME approach has at one point become entirely abandoned, and EPON nowadays implements only the P2PE. The P2PE mode is extended with the addition of the auxiliary single copy broadcast (SCB) port at the OLT. In this

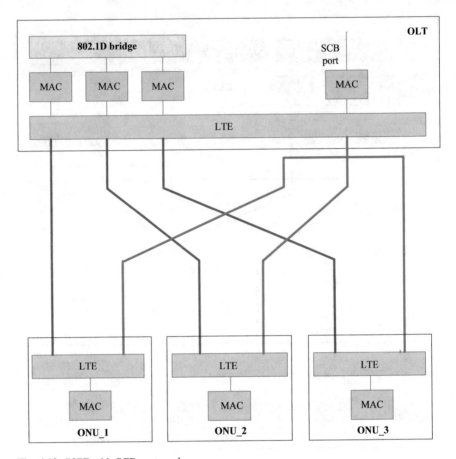

Fig. 4.12 P2PE with SCB port mode

configuration, the OLT contains MAC addresses of all ONUs in the system and one additional MAC address that is used for broadcasting to all ONUs, as shown in Fig. 4.12. In order to optimally separate the traffic, higher sublayers (above MAC) decide to which port the data should be sent. Within this implementation, the SCB channel is only used for the downstream transmission of broadcast traffic. ONUs are not allowed to send upstream frames with defined broadcast LLID [42]. The several special control frames used for ONU's autodiscovery and registration are the exception to this rule.

4.3   EPON Transmission

In accordance with the previous explanations and the functionality of the MPCP, downstream and upstream transmissions are separated and described independently.

4.3.1   Downstream Transmission

In the downstream direction, Ethernet packets are transmitted by the OLT and they pass through a passive splitter or cascade of splitters and reach each ONU. The splitting ratio is limited by the available optical power budget, and it typically ranges from 4 to 64. In this direction, EPON acts as a shared-medium network where its operation coincides with the operation of a conventional 'broadcasting' Ethernet network. Packets are broadcast by the OLT and selectively received by each station based on the destination MAC address from the Ethernet header, as shown in Fig. 4.13.

4.3.2   Upstream Transmission

In the upstream direction, the EPON architecture is thought of as the classical point-to-point architecture where multiple ONUs transmit data packets to the OLT through the common passive combiner and share the same optical fiber from the combiner to the OLT, as shown in Fig. 4.14. In the upstream direction, due to the directional properties of a passive optical splitter/combiner, data packets from any ONU will reach only the OLT and not other ONUs [48, 51–52]. However, all ONUs share the same transmission channel and belong to a single collision domain, i.e., concurrently transmit data over a single optical trunk link. Hence, data packets from different ONUs that are transmitted simultaneously may collide. Consequently, some kind of media access arbitration mechanism has to be deployed in order for data collisions to be avoided and channel capacity among ONUs to be evenly shared.

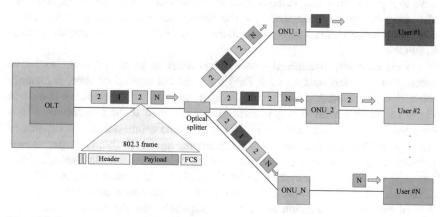

Fig. 4.13 Downstream transmission in EPON

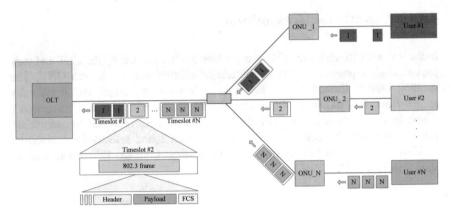

Fig. 4.14 Upstream transmission in EPON

Two possible approaches have been considered in literature: contention-based media access and guaranteed media access. The CSMA/CD mechanism implemented in the half-duplex Ethernet network is an example of contention-based media access mechanism. However, the implementation of the contention-based media access in EPON is extremely difficult because ONUs cannot detect collision due to the directional properties of the optical splitter/combiner. Although the OLT is able to detect a collision and inform the ONUs about it by generating a defined control message, transmission efficiency would ultimately be reduced because of the considerable propagation delay between the OLT and the ONUs. Moreover, with the use of this mechanism, there is no guarantee that a station will get access to the medium even in the short time interval; i.e., the contention-based media access is unable to provide the guaranteed bandwidth to each ONU. Therefore, it is also difficult to support any form of QoS. Having in mind that today's access networks have to support voice and video service in addition to data transmission, and must also provide certain guarantees for the timely delivery of these traffic types, contention-based media access mechanisms are not suitable for the implementation in EPONs.

On the contrary, guaranteed media access schemes grant ONUs the exclusive access to the shared media (trunk link) for a limited interval of time, commonly referred in the literature as the transmission window or timeslot, as shown in Fig. 4.14. In line with this approach, all ONUs have to be synchronized to a common time reference, and each ONU is allocated a timeslot that is capable of carrying several Ethernet packets. Furthermore, an ONU should buffer frames received from users until its timeslot arrives. When the timeslot arrives, ONU transmits buffered frames at a full channel speed and data collisions from different ONUs are avoided. The timeslot for each ONU is allocated in accordance with the defined bandwidth allocation scheme. Consequently, the definition of the bandwidth allocation scheme appears to be one of the key factors for the successful

EPON implementation. In accordance with the timeslot allocation, the bandwidth allocation schemes can be defined as follows:

- the fixed TDMA scheme where the static timeslot allocation is implemented and
- the variable TDMA scheme, i.e., the statistical multiplexing scheme or dynamic bandwidth allocation in which the size of the allocated timeslot is based on instantaneous queue load in every ONU.

Fixed TDMA schemes are easier to implement since each ONU can be configured to start and stop data transmission at the predetermined repeating intervals. On the other hand, in case of its implementation, the system efficiency is very low due to the presence of bursty data or variable-size packets. In fixed schemes, a predefined timeslot is allocated to every ONU, regardless of its current buffer occupancy, which can be either empty or much less occupied in comparison with the size of the allocated window. On the other hand, it may result in a situation in which some timeslots overflow even under very light load, causing packets to be delayed for several timeslots.

For the reasons stated above, the current infield EPON implementation includes the implementation of dynamic bandwidth allocation schemes. In order to increase utilization and system efficiency, the OLT dynamically allocates a variable timeslot to each ONU which is based on the instantaneous bandwidth demand of the ONUs. These schemes can be further divided in two groups:

- centralized bandwidth allocation schemes and
- distributed bandwidth allocation schemes.

In the latter approach, each ONU decides when and how much buffered data should be sent to the OLT. Although these schemes avoid the waste of bandwidth, they have one major drawback. Namely, in order to avoid collisions between stations, some kind of communication between the ONUs has to be enabled. Accordingly, the network should be deployed as a ring or as a broadcasting star. This does not appear to be a cost-effective solution since, in comparison with the tree topology, i.e., point-to-multipoint topology, the deployment of more fiber is required.

The centralized approach is prevalent in today's EPON realizations, and here, the OLT remains the only device that can arbitrate the time-division access to the shared channel. However, in accordance with this approach, the OLT does not know how many bytes of data exist in the each ONU buffer even though this is necessary for the accurate timeslot assignment. The simplest approach is to implement the request/grant mechanism in which every ONU reports its current buffer occupancy by means of generating the request message. The OLT processes all requests, and based on the available bandwidth, it allocates different transmission windows to ONUs by generating grant messages. The main advantage of the centralized intelligence is that the OLT knows the state of the entire network and can switch to another allocation scheme based on that information.

However, the bandwidth allocation algorithms may depend on many parameters, among which are the deployment conditions, the supported application, and services and the defined SLAs. Consequently, the IEEE 802.3ah task force has decided to standardize only the control and management messages used to control the data exchange between the OLT and the ONUs as well as the processing of these messages through the development of the MPCP. The group decided to leave the definition of the dynamic bandwidth allocation algorithm out of the standard. The DBA implementation is left to the equipment vendors, where the device interoperability is ensured by the standardization of the MPCP. The MPCP is not concerned with a specific DBA algorithm; it is just a supporting control mechanism that facilitates the implementation of various bandwidth allocation schemes in EPON.

4.4 Dynamic Bandwidth Allocation in EPON

In this section, we focus on the communication and data transmission in EPON. In order to do that, we have to define the term 'transmission cycle' in the first place. The transmission cycle is defined as a time interval between the moment when the foremost GATE message is completely received and the moment when the lattermost REPORT message is completely transmitted. In order to separate transmissions of different ONUs, a guard time is scheduled before each transmission window to perform the ranging for different distances between the OLT and ONUs and to provide transmitter switching time.

As previously explained, the DBA algorithm in EPON relies on two Ethernet control messages (GATE and REPORT) in its regular operation. In the one transmission cycle, every ONU must get the chance to transmit a portion of buffered frames and exchange the control messages with the OLT. Within each cycle and allocated window, the ONU sends REPORT message along with data in order to report bandwidth requirements for the upstream transmission of its traffic. A REPORT message can be either transmitted at the beginning of the timeslot, or at the end, depending on the bandwidth request approach implemented by the ONU. It contains the requested size of the next timeslot based on the buffer occupancy of the ONUs [47].

Each ONU uses a set of queues to store its frames and starts transmitting them as soon as its transmission window starts. The intra-ONU scheduler schedules the packet transmission for various traffic queues from local users, and the transmission window may comprise multiple Ethernet frames. An ONU can support up to 8 priority queues, as defined in 802.1Q [53]. The ONU should also account for additional overhead when requesting the next timeslot that includes an 8-byte frame preamble and a 12-byte inter-frame gap (IFG) between two consecutive frames. Between the allocated transmission windows of two ONUs, a certain guard time is needed to account for the laser on and off times, receiver recovery times, round-trip delay (which relates to the physical distance between the communicating ONUs), and other optic-related issues.

Upon receiving REPORT messages from all ONUs in the defined cycle, the OLT passes the messages to the DBA module which performs the bandwidth allocation computation. Apart from that, the DBA module will recalculate the RTT for the each ONU. The OLT assigns the transmission windows (TWs) via GATE messages, as shown in Fig. 4.15. The GATE message specifies the transmission start time and end time during which each ONU can transmit the queued customer traffic upstream to the OLT. The transmission could include multiple Ethernet frames, depending on the size of the allocated transmission window and the number of buffered packets at the ONU. Since packet fragmentation is allowed, in case the next frame does not fit the allocated timeslot, it will be deferred to the next timeslot [51–52].

4.4.1 Review of the DBA Algorithms

According to the IEEE 802.3ah standard, the size and periodicity of the discovery windows, i.e., the definition of the bandwidth allocation algorithms, are left to vendors. So far, various bandwidth allocation schemes have been proposed. Here, we focus on and review the most influential ones and discuss their characteristics and implementation. The bandwidth allocation schemes could be grouped in two categories: static and dynamic bandwidth allocation schemes.

A. *IPACT description*

Authors in [54] have proposed an OLT-based polling algorithm called IPACT (interleaved polling with adaptive cycle time). IPACT requires the OLT to poll every ONU and dynamically assign to it bandwidth before transmission. In this

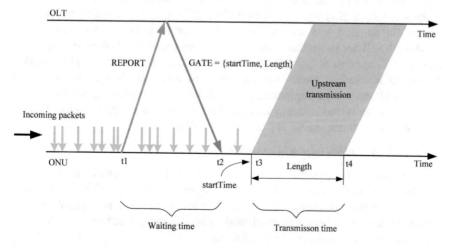

Fig. 4.15 REPORT/GATE mechanisms and upstream transmission

approach, the OLT polls every ONU individually and issues a transmission grant in a round-robin fashion. Upon the completion of the discovery procedure, the OLT issues individual GATE messages to all discovered ONUs. These GATE messages contain only one grant and require the ONUs to send REPORT messages in the corresponding timeslots. The timeslot length in each GATE message is sufficient for sending only one MPCPDU. When a REPORT message from the first ONU arrives, the OLT allocates a timeslot. If the ONU buffer is empty, it will report 0 bytes back to the OLT. Correspondingly, in the next cycle, this ONU will be granted a small timeslot sufficient for sending only a REPORT message, but no data. After the first cycle, the OLT's receiving channel is almost 100% utilized (REPORT messages and guard times consume some bandwidth). Idle ONUs (those without data to send) are given very short transmission windows. This leads to a shortened cycle time, which, in turn, results in more frequent polling of active ONUs.

Moreover, IPACT deploys an in-band signaling of bandwidth requests by using escape characters within Ethernet frames instead of using an entire Ethernet frame for control of bandwidth requests in the MPCP. Consequently, this results in a reduced signaling overhead. The OLT keeps track of the RTTs of all ONUs, and hence, the OLT can send out a grant to the next ONU in order to achieve a very tight guard band between consecutive upstream transmissions, resulting in improved bandwidth utilization. The guard band between two consecutive upstream transmissions is needed to compensate RTT fluctuations and to give the OLT enough time to adjust its receiver to the transmission power level of the next ONU. In IPACT, each ONU is served once per round-robin polling cycle. The cycle length is not static but adapts to the instantaneous bandwidth requirements of the ONUs. By using a maximum transmission window (MTW), ONUs with high traffic volume are prevented from monopolizing the bandwidth. Therefore, this approach allows the OLT to send a GATE message to the next ONU before the data and REPORT message(s) from the previous polled ONU(s) arrive, as shown in Fig. 4.16. Since the upstream and downstream channels are separated, the OLT maintains information about each ONU in a polling table, including the reported bandwidth demand and the calculated RTT for each ONU. Any unrequested bandwidth will not be granted, and the scheduling frame size is therefore not fixed. In the OLT, upon receiving REPORT messages from all ONUs, the DBA module begins to recalculate bandwidth for the each end-user. Hence, the presented approach allows for the implementation of different bandwidth algorithms, and the authors have provided in-depth study of the following ones: fixed, gated, limited, constant credit, linear credit, and elastic algorithm.

B. *Fixed bandwidth allocation scheme*

Fixed algorithm ignores the requested window and always grants the maximum window size (static timeslot) to each ONU in every cycle regardless of its actual needs, i.e., buffer occupancy. This scheme is simple to implement and essentially works as the previously described static slot assignment scheme in which the timeslots allocated to every ONU are fixed.

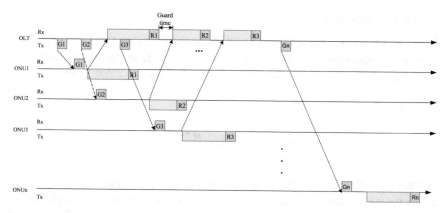

Fig. 4.16 IPACT algorithm

Accordingly, the allocated bandwidth is calculated as follows:

$$W_i^{\text{allocated}} = W^{\text{max}}. \tag{4.3}$$

where $W_i^{\text{allocated}}$ presents the bandwidth allocated to ONU$_i$ in the observed transmission cycle, and W^{max} presents the maximum transmission window.

The main drawback of this model lies in the fact that one ONU will occupy the upstream channel for its assigned timeslot even if there is no frame to transmit. As a result, the delay would increase for all Ethernet frames buffered in other ONUs. Moreover, light-loaded ONUs will probably underutilize their allocated slots, which would lead to the increased delay of other ONUs. Eventually, the throughput of the system will be significantly degraded.

As a result of all of the above stated, this approach has been abandoned and the dynamic schemes have became prevalent. The authors have studied the performance of EPON using a dynamic bandwidth assignment algorithm in cases where all traffic belonged to a single class, i.e., without service differentiation.

C. *Gated algorithm*

In this scheme, the OLT always authorizes an ONU to send as much data as it has requested. In order to limit the duration of the transmission cycle, the authors introduced the buffer size (W^{queue}) as the limiting parameter. Namely, since an ONU cannot store more than W^{queue} bytes, it will never request more than W^{queue} bytes. Hence, the bandwidth allocated for ONU$_i$ in one transmission cycle is calculated as follows:

$$W_i^{\text{allocated}} = W_i^{\text{requested}}. \tag{4.4}$$

where $W_i^{\text{requested}}$ presents the request bandwidth of ONU$_i$ in the observed transmission cycle.

D. *Limited algorithm*

In the limited allocation scheme, the OLT simply grants an ONU the number of bytes it has requested, not exceeding the maximum size of the transmission window. This algorithm has the shortest cycle of all of the presented schemes.

The bandwidth allocated to ONU_i in one transmission cycle is defined as follows:

$$W_i^{\text{allocated}} = \min \begin{cases} W_i^{\text{requested}} \\ W^{\text{max}} \end{cases}. \qquad (4.5)$$

The timeslot allocated to ONU is limited by the maximum time (maximum size of the transmission window) allowed for an ONU, i.e., the maximal slot length (W^{ma}). W^{ma} parameter can be specified by SLA or other system parameters. The granted timeslot in this implementation may vary in accordance with the dynamic traffic, and the duration of the service cycle also varies due to the fact that ONUs are assigned different timeslot lengths in different service cycles.

This is the most conservative scheme because it assumes that no more packets had arrived after the ONU sent its request. However, because of the RTT between the OLT and each ONU, there might be more packets arriving in the waiting time interval, i.e., between the moment when an ONU sends a REPORT message and the moment when the ONU receives a GATE message. Those newly arriving packets will most probably not be transmitted in the current cycle, resulting in the increased average packet delay. In order to address this problem, the constant credit and linear credit schemes were proposed.

E. *Constant credit algorithm*

Under the REPORT/GATE mechanisms, when sending a REPORT message, an ONU will only report the already buffered frames to the OLT. Therefore, frames that arrive during the waiting time have to be deferred to the next timeslot (Fig. 4.15) even if the upstream channel is lightly loaded. Constant credit scheme takes such frames into consideration and adds a constant credit to the requested window size.

The size of the credit is constant no matter how large the requested window size may be. Once an ONU receives a GATE message, it can send packets amounting to the requested window size plus the constant credit in the following way:

$$W_i^{\text{allocated}} = \min \begin{cases} W_i^{\text{requested}} \\ W^{\text{max}} \end{cases} + C, \qquad (4.6)$$

where C presents the credit. C could be a constant or linear credit. The frames buffered during the waiting time interval are expected to be transmitted (or partially transmitted) within the current timeslot.

In practice, the choice of credit size may have an impact on network performance. A too small size will not be able to significantly improve the packet delay.

On the other side, an exceedingly large size will reduce the bandwidth utilization of the upstream channel. The choice should be based on traffic, i.e., the application characteristics or some other empirical data.

F. *Linear credit algorithm*

Linear credit scheme has the similar approach as the one used in the constant credit scheme, but in this approach, the size of the credit is proportional to the requested window. The basis of this scheme is that network traffic usually has a certain degree of predictability. This means that if a long burst of data is observed, this burst is very likely to continue for a longer time; i.e., long burst of data is in most cases followed by the traffic with the bursty characteristics as well.

The size of the allocated window could be calculated as follows:

$$W_i^{\text{allocated}} = \min \begin{cases} W_i^{\text{requested}} \cdot C \\ W^{\text{max}} \end{cases}, \tag{4.7}$$

G. *The elastic algorithm*

The elastic service avoids the definition of the fixed maximum window size and uses the maximum cycle time T_{max} as the only limiting factor. The maximum window is granted in such a way that the accumulated size of the last $i \leq N$ grants (including the one being granted) does not exceed $N \cdot W^{\text{max}}$, where N presents the number of ONUs in the system:

$$W_i^{\text{allocated}} = \min \begin{cases} W_i^{\text{requested}} \\ N \cdot W^{\text{max}} - \sum_{k=1}^{i-1} W_k^{\text{allocated}} \end{cases}. \tag{4.8}$$

The results obtained in [54] indicate that the interleaved polling protocol can significantly improve network performance in terms of channel utilization and average packet delay. Among these algorithms, the limited scheme exhibits the best performance. However, although this scheme provides statistical multiplexing and results in efficient channel utilization, the algorithm is not suitable for delay and jitter-sensitive services because of the variable polling cycle time. Moreover, this protocol allows the OLT to allocate bandwidth which is based solely on the already received bandwidth demands.

H. *Excessive bandwidth reallocation*

Besides the IPACT model and its flavors, one of the dominant bandwidth allocation models is presented in [55] and it includes the summarization of the underexploited bandwidth of lightly loaded ONUs (excessive bandwidth) and its redistribution in the system among the heavily loaded ONUs.

As we have previously explained, the limited IPACT scheme grants the requested number of bytes to an ONU but not more than the predefined value called the minimum guaranteed bandwidth. However, due to the busty nature of Ethernet traffic, some ONUs might have less traffic to transmit, while other ONUs require more than the minimum guaranteed bandwidth. To improve the limited bandwidth allocation algorithm, authors of [55] exploit excessive bandwidth (saved from under-loaded ONUs) by fairly distributing it among the highly loaded ONUs.

In the presented scheme, called weighted inter-ONU DBA algorithm, all ONUs could be partitioned into two groups—the under-loaded and the overloaded ONUs, based on their requested bandwidth. ONUs that request less bandwidth than the minimum guaranteed are in the first group, while the ONUs that require more bandwidth than the minimum guaranteed belong to an overloaded group. The sum of the underexploited bandwidth of lightly loaded ONUs is called excessive bandwidth. In the proposed solution, the total bandwidth saved from the under-loaded ONUs is reallocated to an overloaded group in order to improve system efficiency.

The total available bandwidth in one granting cycle can be expressed as follows:

$$B^{\text{total}} = R \cdot \left(T_{\text{cycle}}^{\text{max}} - T_{\text{g}} \right), \qquad (4.9)$$

where R presents the OLT link capacity, $T_{\text{cycle}}^{\text{max}}$ denotes the maximum transmission cycle time (MTCT), and T_{g} denotes the guard time interval. The granting cycle is the time interval during which all active ONUs can transmit and/or report to the OLT. The duration of the granting cycle is very important for the efficient transmission and high bandwidth utilization. If the granting cycle is too large, that results in the larger transmission size and in case of light-loaded ONUs, it will be underutilized. On the other hand, in case of heavy-loaded ONUs, the large granting cycle could result in lower average packet delays. However, maximum packet delays will be increased due to the increased load of the transmission media. If the granting cycle is too small, it will result in the decreased bandwidth efficiency since more bandwidth will be wasted by guard intervals. The guard interval, however, must be taken into account in order to provide protection for the fluctuations in RTT of different ONUs. Consequently, the CPU processing load will increase, and large packets will most probably be lost because packet fragmentation is not allowed.

The minimum guaranteed bandwidth for ONU, i.e., the minimum bandwidth which the OLT allocates to ONU under heavy-load operation (i.e., peak times), can be calculated as follows:

$$B_i^{\text{min}} = \frac{\left(T_{\text{cycle}} - N \cdot T_{\text{g}} \right) \cdot R}{8} \cdot w_i. \qquad (4.10)$$

where w_i presents the weight factor assigned to $\text{ONU}_i \left(\sum_{i=1}^{N} w_i = 1 \right)$. Under assumption that all ONUs need to be classified on their SLA ($w_i = w = 1/N$), the total minimum guaranteed bandwidth for each ONU will be:

$$B_i^{min} = \frac{(T_{cycle} - N \cdot T_g) \cdot R}{8 \cdot N}. \tag{4.11}$$

According to the limited bandwidth allocation scheme (Sect. 4.4.1 (D) and formula (4.5)), the bandwidth allocated to ONU is as follows:

$$B_i^{allocated} = \begin{cases} B_i^{requested}, & \text{if } B_i^{requested} < B_i^{min} \\ B_i^{min}, & \text{if } B_i^{requested} \geq B_i^{min} \end{cases} . \tag{4.12}$$

Furthermore, the total excessive bandwidth saved by M under-loaded ONUs and the total extra bandwidth demanded by K overloaded ONUs, in the observed cycle, are given as follows:

$$B_{total}^{excess} = \sum_{i=1}^{M} \left(B_i^{min} - B_i^{requested} \right), \quad B_i^{min} > B_i^{requested} \tag{4.13}$$

In order to improve the efficiency of the limited bandwidth allocation scheme, the total excessive bandwidth is further fairly redistributed to overloaded ONUs in the following way:

$$B_i^{excess} = \frac{B_{total}^{excess} \times B_i^{requested}}{\sum_{i=1}^{K} B_i^{requested}}, \tag{4.14}$$

$$B_i^{allocated} = B_i^{min} + B_i^{excess}, \tag{4.15}$$

where B_i^{excess} is the excessive bandwidth allocated to ONU$_i$.

The presented algorithm can allocate bandwidth more efficiently than the limited scheme. However, the presented approach and algorithm assume that the saved bandwidth will always be fully occupied by overloaded ONUs, which is not necessarily true. Consequently, that would result in the decreased bandwidth efficiency due to the wasted capacity, which, in effect, constitutes the main drawback of this algorithm.

4.5 QoS Support in EPON

Although cost-effective, the EPON architecture is bandwidth limited and does not scale well with the increased number of users and ONUs. Accordingly, QoS support has become the key concern for the successful implementation of EPONs [56–58].

QoS implementation and multiservice provisioning in EPONs include the traffic classification mechanism and scheduling algorithms. Namely, for the purpose of prioritizing traffic and implementing QoS mechanisms, EPONs have to give different priority to different applications, users, or data flows, as described in the third chapter. Hence, EPONs must be able to classify traffic into service classes and provide the differentiated treatment of each class [48]. Beside the traffic classification in supporting differentiated QoS, the scheduling algorithms have to be implemented as well. There are two types of scheduling paradigms: inter-ONU and intra-ONU scheduling [59].

Inter-ONU scheduling is responsible for arbitrating the transmissions of different ONUs, and intra-ONU scheduling is responsible for arbitrating the transmissions of different priority queues in each ONU. There are two approaches for the implementation of these scheduling paradigms:

- The OLT performs both inter-ONU and intra-ONU scheduling. The OLT is the only device that arbitrates upstream transmissions. In this case, the OLT allocates bandwidth to each traffic class in every ONU based on the previously reported bandwidth requests. Hence, each ONU must report the status of its individual priority queues to the OLT through REPORT messages. MPCP specifies that each ONU can report the status of up to eight priority queues [42]. The OLT can then generate multiple grants, each for a specific traffic class, to be sent to the ONU using a single GATE message and
- The OLT performs inter-ONU scheduling and allows each ONU to perform intra-ONU scheduling. In this case, each ONU requests the OLT to allocate bandwidth to it based on its buffer occupancy status. The OLT only allocates the requested bandwidth to each ONU. Further, each ONU will divide the allocated bandwidth among different classes of services based on their QoS requirements and schedule the transmissions of different priority queues within the allocated bandwidth. There are two types of scheduling algorithms: strict and non-strict priority scheduling. These algorithms will be explained in detail in the following sections.

In addition to the DiffServ support, we implement the second approach, and consequently, we discuss the implementation of the two independent scheduling mechanisms:

- Inter-ONU: scheduling at the OLT (inter-ONU scheduling or dynamic bandwidth allocation (DBA)) and
- Intra-ONU: scheduling at the ONU for scheduling packets that belong to different traffic classes.

In general, the combination of both scheduling mechanisms is required to enable the centralized EPON networking architectures to support QoS. Figure 4.17 shows the interaction between the algorithms for scheduling: An algorithm for inter-ONU scheduling defines the amount of bandwidth allocated to each ONU in the system for the upstream transmission, while the algorithm for intra-ONU scheduling defines how packets, with different priorities and from different queues, are allocated for transmission within the allocated transmission window.

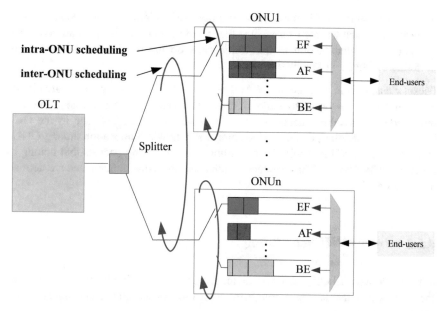

Fig. 4.17 Scheduling in EPON

In the described system, the OLT is the only device that can arbitrate the time-division access to the shared channel. Since the OLT has global knowledge of the state of the entire network, this is a control plane in which the OLT has centralized intelligence. This approach is currently prevalent in the literature, although there are solutions for the decentralized EPON architecture [45]. However, in the decentralized EPON architecture, some of the functionalities of the OLT unit are transferred to the ONUs which rapidly increases their price, and with it the price of the whole system as well, which further questions the feasibility of such a system.

4.5.1 Traffic Classification

For multiservice provisioning in EPONs, the network traffic should be categorized into three different classes which are consistent with the DiffServ framework [60]:

- EF (expedited forwarding) traffic class—highest priority traffic class for delay-sensitive traffic with constant bit rate. This class is intended for services such as voice and other delay-sensitive applications that require bounded end-to-end delay and jitter specifications;
- AF (assured forwarding) traffic class—medium-priority traffic for not delay-sensitive traffic with variable bit rate. AF class is intended for services such as video transmission that are not delay-sensitive but which require bandwidth guarantees; and

- BE (best-effort) traffic class—low priority traffic class for delay-tolerable services that include Web browsing, file transfer, and e-mail applications. Also, applications that belong to this traffic class do not require any guarantees in terms of jitter and bandwidth.

Packet classification is based on the ToS field of every IP packet encapsulated in the Ethernet frame. In order to support different services, it is necessary that two independent scheduling mechanisms be implemented in EPON, one in the OLT (inter-ONU scheduling or DBA algorithm) for bandwidth allocation among ONUs and one in the ONU (intra-ONU scheduling) for scheduling packets that belong to different traffic classes. Three queues which share the same buffer space are defined in every ONU.

4.5.2 Inter-ONU Scheduling

In the centralized EPON, OLT is a device which is responsible for the arbitration of the upstream transmissions through the allocation of an appropriate transmission window to each ONU.

Since the MPCP does not specify any bandwidth allocation algorithm, an independent algorithm must be deployed at the OLT to assign a portion of the available upstream bandwidth to every ONU.

A key factor for the realization of the efficient QoS-based EPON is the bandwidth allocation algorithm. The main goal of bandwidth allocation is to effectively and efficiently perform the fair scheduling of timeslots between the ONUs in EPONs. As explained in Sect. 4.4, each ONU can transmit packets only during its assigned transmission window. Hence, within an allocated window, each ONU periodically reports its buffer occupancy status to the OLT through the generation of the REPORT message and requests slot allocation. Upon receiving the message, the OLT passes this information to the DBA module, as shown in Fig. 4.18.

Based on a REPORT message that contains queue occupancy information, an inter-ONU scheduler (DBA module) in the OLT allocates the appropriate bandwidth. The grant allocation table is updated by the output of the DBA algorithm. Grant instructions are then compiled into MPCP GATE messages and transmitted to the ONUs following the performance of the RTT compensation. The GATE message (each GATE may carry more than one grant message) specifies the transmission start time and end time during which the ONU can transmit queued customer traffic upstream to the OLT, as shown in Fig. 4.19.

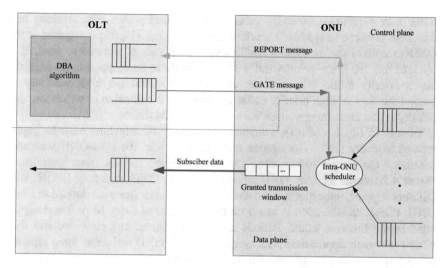

Fig. 4.18 Exchange of control messages

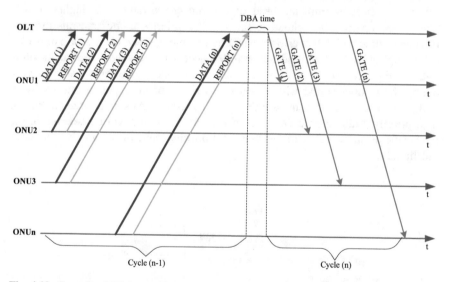

Fig. 4.19 Centralized DBA operation

4.5.3 Intra-ONU Scheduling

The queuing discipline defines how the system treats packets with different priorities. In case an arriving packet with higher priority finds the buffer full, it can displace a lower-priority packet. Otherwise, if a low priority packet arrives and the buffer is full, then the low priority packet is dropped. In this way, lower-priority traffic may experience excessive delays and increased packet loss, resulting in a complete resource starvation. Therefore, traffic policing [54, 59] is required at the ONU to control the amount of traffic which each user is allowed to send. It is obvious that the lower-priority traffic is more likely to be dropped in favor of the higher-priority traffic. Nevertheless, control mechanisms are also necessary to control the flow of high priority traffic and to avoid the situation in which the high priority traffic class monopolizes the available bandwidth.

In case of EPONs with QoS support, the Intra-ONU scheduler must be implemented in every ONU. During one transmission cycle, the intra-ONU scheduler schedules the packet transmission for various traffic queues from local users. Figure 4.20 shows the Intra-ONU scheduler in every ONU. In case of EPONs with DiffServ support, three queues who share the same buffer space are defined in every ONU. Packet classification is based on the ToS field of every IP packet encapsulated in the Ethernet frame. Packets are first segregated and classified and then placed into their appropriate priority queues. Each ONU maintains three separate priority queues for EF, AF, and BE traffic classes.

The strict priority scheduling mechanisms are defined as default Intra-ONU scheduling mechanisms in the single-channel EPONs [42, 48]. Strict priority scheduler schedules from the head of a given queue only if all higher-priority queues are empty. This situation favors high priority traffic transmission and penalizes traffic with lower priority resulting in, among other, the increase of packet delay, higher packet loss, and uncontrolled access to the shared media. As illustrated in Fig. 4.15, in the strict priority scheduling, the high priority traffic arriving during the waiting period will be scheduled ahead of the reported lower-priority traffic. As a result of this, in case the system needs to transmit a large amount of high priority traffic, the transmission of low priority traffic will be deferred to the next cycle or more cycles. In this way, low priority traffic queuing delay increases indefinitely.

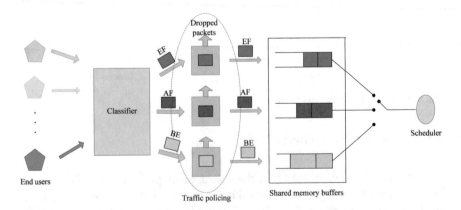

Fig. 4.20 QoS mechanisms

4.6 Algorithms for QoS Support in EPON

Until today, a lot of research has been done on this subject. Currently, most attention in the literature is given to algorithms for dynamic bandwidth allocation with quality of service support, i.e., algorithms that include both inter- and intra-ONU scheduling mechanisms. We briefly review the main characteristics of the most influential models in Sects. 4.6.1–4.6.3. In Sect. 4.6.4, we present the hybrid granting algorithm with priority-based scheduling (HG(PBS)) that tends to be more efficient in comparison with other algorithms which have been proposed until now and introduce both the inter- and intra-ONU scheduling in EPONs [48]

4.6.1 DBA for Multimedia Services Over Ethernet PONs

The DBA algorithm for the transmission of multimedia services over EPONs is suggested in [61]. In the presented scheme, traffic is classified and placed into one of the three priority queues—high, medium, or low. Further, a REPORT/GRANT mechanism is used for resource allocation. Each ONU informs (through REPORT massages) the OLT about the sizes of all three priority queues. Based on such information, the OLT issues grants separately for each of the priority classes in each ONU where these priority classes are used by the DBA algorithm for bandwidth assignment.

The DBA algorithm works as follows. Bandwidth is first handed out to the high priority queues, satisfying all the requests of high priority traffic flows. After that, the DBA algorithm considers the requests from the medium-priority traffic flows. If the remaining bandwidth can satisfy all of the medium-priority flow requests with what is left over from high priority traffic, the remaining bandwidth is then allocated to each ONU for the transmission of the medium-priority traffic flows. If the requested bandwidth for the medium-priority traffic transmission is larger than what is left over from high priority traffic (available bandwidth), the suggested DBA divides the remaining bandwidth between all medium-priority flows. The fraction of the bandwidth granted to each medium-priority flow is related to the amount of bandwidth requested by each flow of the total of all medium-priority requests. Following the medium-priority traffic processing, the remaining bandwidth, if any, is distributed among low priority flows in the same way as for medium-priority traffic.

In the presented approach, bandwidth is essentially allocated using strict priority scheduling based on the requirements of each priority traffic class of the entire EPON (all ONUs are connected to a single OLT). Moreover, the OLT controls the scheduling within the ONU; i.e., each ONU reports the occupancy of each of the three priority queues and the OLT must issue multiple grants to each ONU per cycle. Consequently, OLT processing becomes more complicated, and overall system complexity increases. Moreover, the implemented strict priority scheduling based on traffic classes at the PON level may result in the starvation of lower-priority traffic that will ultimately lead to a decrease in system efficiency.

4.6.2 IPACT Extension to Multiple Service Classes

The IPACT extension to multiple service classes presented in [62] may be viewed as a precursor to the DBA for QoS approach presented in [55]. The presented algorithm includes support for differentiated classes of service and the implementation of three traffic classes with strict priority scheduling inside the ONU. This is similar to the previously described allocation scheme, but while, in the previously presented DBA for multimedia services transmission multimedia priority scheduling was performed at the EPON level (all ONUs connected to a single OLT), in the presented IPACT extension, priority scheduling is performed at the ONU level.

The authors have noticed and analyzed a phenomenon they call 'light-load penalty.' Namely, they have noticed that under light loading, the lower-priority traffic class experiences a significant increase in the average packet delay. Moreover, the maximum packet delays for the higher priorities also exhibited a similar behavior. The reason for this lies in the fact that queue reporting occurs at some point before the strict priority scheduling is performed. This allows the higher-priority traffic, which arrives after queue reporting but before the transmission grant, to preempt lower-priority traffic that had arrived before queue reporting. At light load, this problem accelerates and significantly degrades system efficiency.

The authors further discuss two methods for resolving this issue. In the first approach, packets are scheduled when the REPORT message is transmitted and placed in the second-stage queue. When the GATE message from the OLT arrives, the second-stage queue will be emptied out first into the provided timeslot. The main drawback of this method is the increase in the average packet delay of the highest priority class which goes beyond one cycle time. In order to resolve that issue, and as part of the second method, the authors use the fact that high priority traffic has constant bit rate (CBR) characteristics with a given data rate. Accordingly, authors suggest the possibility of predicting the number of high priority packets that arrive in the observed ONU between queue reporting (REPORT messages transmission) and the grant window (GATE message arrival) so that the grant window will be large enough to accommodate the newly arriving high priority packets. This method inherently lowers the delay experienced by higher-priority traffic as opposed to the delay experienced by higher-priority traffic in the two-stage queuing approach.

4.6.3 DBA for the QoS Model

In the DBA for the QoS model [55], the authors incorporate a method similar to the two-stage queuing approach explained above. Specifically, in the DBA for the QoS method, the packet scheduler in the ONU employs priority scheduling only on the packets that arrive before the time t_1 at which the REPORT message is sent to the OLT, as shown in Fig. 4.15. This avoids the problem caused by the fact that the ONU packet scheduler requests bandwidth based on buffer occupancies at time t_1 and then actually schedules packets at time t_2 (GATE message arrival) to fill the granted transmission window.

As we have previously explained, without this mechanism, the lower-priority traffic could be delayed because the higher-priority traffic that arrives during the waiting time would take away the transmission capacity assigned to the lower-priority queues. However, this problem only arises with strict priority scheduling, which schedules lower-priority packets only when the higher-priority packet queues are empty. In case of the implementation of a more advanced queuing mechanism, this problem would not arise. The authors of [55] suggest the implementation of the weighted fair queuing, which serves different priority queues in proportion to fixed weights. As a result, in the proposed DBA for the QoS scheme, each ONU is assigned a guaranteed bandwidth in proportion to its SLA; i.e., the total upstream bandwidth is divided among the ONUs in proportion to their SLAs.

The implemented DBA algorithm takes into account the excess bandwidth; i.e., the unused bandwidth of the light-loaded ONUs (ONUs the requested bandwidth of which is smaller than their guaranteed bandwidth) is distributed to each of the highly loaded ONUs (ONUs the requested bandwidth of which is larger than their guaranteed bandwidth) in a manner that weighs the excess assigned in proportion to the size of their request. The presented proportional scheduling approach is in contrast with the strict priority scheduling of the DBA for multimedia, which does not allocate any bandwidth to lower-priority traffic classes until the bandwidth demand of all higher-priority traffic classes is met. Moreover, the DBA for the QoS allows for the sending of the individual priority queue occupancies of each ONU to the OLT via REPORT messages (a REPORT message supports reporting queue sizes of up to eight queues), and here, the OLT generates transmission windows for the each individual priority queue (the GATE message supports the sending of up to four transmission grants). This option puts priority scheduling, which would otherwise be handled by the ONU, under the control of the OLT.

The DBA for the QoS also considers the option of reporting the queue size by means of using an estimator of the occupancy of the high priority queue. As mentioned earlier, packets that arrive during the waiting time will have their transmission deferred to the cycle after the next one, which will highly likely pose additional delays. In practice, some applications might tolerate this, but the delay-sensitive traffic will not. To prevent the high priority traffic from being penalized, the author suggests a solution in which the ONU estimates (based on the certain statistical history from the previous cycles) the bandwidth required by this type of traffic arriving during the waiting time.

4.6.4 Hybrid Granting Protocol with Priority-Based Scheduling (HG(PBS))

The HG(PBS) model [48] separates the transmission of high priority traffic from the transmission of lower-priority traffic and introduces the implementation of Intra-ONU scheduling algorithms for the transmission of lower traffic class. Bearing in mind the announced dominance of multimedia traffic in the access network, as well as the rapid development of various applications characterized by variable bit rate, i.e., the class of medium-priority traffic, guaranteeing the QoS for this class becomes a key issue for the successful implementation of a multiservice EPON.

In the presented solution, the dynamic bandwidth allocation is based on the HG (hybrid grant) protocol [63] that divides the transmission cycle into two sub-cycles for data transmission, i.e., EF subcycle and AF/BE subcycle. In standard EPON algorithms, as explained in Sect. 4.4 and presented in Fig. 4.15, the MPCP is implemented in a GAR (GRANT after REPORT) fashion. In this mechanism, the DBA module in the OLT generates a GATE message with the allocated bandwidth for ONU following the receipt of a REPORT message from that ONU with its requested bandwidth. The implementation of a GAR mechanism allows for defining the minimal queuing delay in every ONU, but this does not represent an optimal solution for the transmission of traffic that is not delay-sensitive.

The HG protocol takes into account the facts that the amount of the EF traffic in the system is fully deterministic [64] and accordingly introduces the GBR (GRANT before REPORT) mechanisms. In GBR mechanisms, a GATE message transmits information about the expected EF traffic that will arrive in the system prior to the next transmission cycle in the given ONU. Hence, it is possible to define the maximum queuing time for EF packets. Since bandwidth allocation for EF traffic is not based on the last received REPORT message, the OLT has to allocate bandwidth for the EF traffic in advance, prior to allocating bandwidth for lower traffic classes.

Hence, the HG protocol defines two subcycles for data transmission, one for the EF traffic, with the use of a GBR mechanism, and one for the AF/BE traffic, with the use of a GAR mechanism. Since bandwidth allocation for EF traffic is not based on the last received REPORT message, i.e., the OLT has to allocate bandwidth for EF traffic in advance, before allocating bandwidth for the rest of the traffic in the system, the OLT has to predict the precise beginning of the next cycle in every ONU. The bandwidth for EF traffic is always allocated before the AF/BE subcycle. The remaining bandwidth is used for the transmission of AF/BE traffic, taking into account the maximal bandwidth defined by the MTCT. Since the behavior of AF and BE traffic is non-deterministic and is characterized with variable bit rate transmission, for the transmission of these types of traffic the authors implement the standard GAR technique and a DBA algorithm with QoS support. The MTCT parameter defines the maximum transmission interval in ONU, called the DBA cycle. Based on the MTCT, the minimal guaranteed bandwidth for every ONU in one cycle can be defined as the sum of EF and AF/BE subcycles.

Apart form the GBR mechanisms, authors in [63] have introduced two new concepts, namely the DBA cycle (DBA-CL) and the MPCP cycle (MPCP-CL). The DBA-CL is computation-based and MPCP-CL is operational-based. The DBA-CL is defined by the MTCT parameter and determines the minimum guaranteed bandwidth for each ONU in one cycle. The MPCP-CL defines the exchange of MPCP messages between the OLT and every ONU.

In the standard DBA algorithms, these cycles are co-phased, i.e., both the DBA-CL and MPCP-CL are completely overlapped within a scheduling frame (the EF traffic class has the highest, and hence, the EF subcycle in the DBA-CL is always scheduled ahead of the AF subcycle), as shown in Fig. 4.21a. However, the definition of the GBR mechanisms allows for the shifting of the MPCP-CL in time.

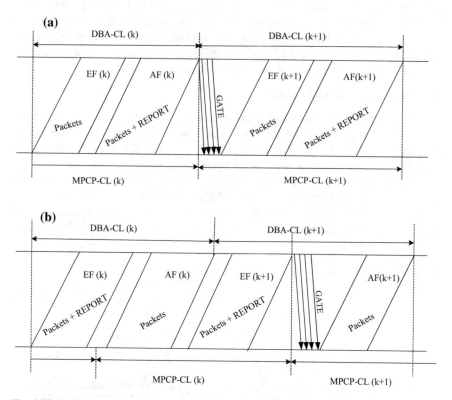

Fig. 4.21 Definition of the DBA-CL and MPCP-CL cycles **a** GAR mechanisms; **b** GBR mechanisms

Since the window size for the EF subcycle is allocated in the GBR fashion and the AF subcycle in the GAR fashion, it is not necessary for the ONU to request bandwidth for EF packets. At the end of its AF transmission window, each ONU sends a REPORT message to the OLT to request bandwidth for its buffered AF and BE packets. Now, the MPCP-CL begins with an AF subcycle (k) and ends with an EF subcycle ($k + 1$), while the DBA-CL still begins with an EF subcycle (k) and ends with an AF subcycle (k), as shown in Fig. 4.21b. In this way, the ONU sends the REPORT message at the end of the each EF transmission window with the information about the allocated bandwidth for the lower-priority classes in the current subcycle and the information about the allocated bandwidth for the transmission of the EF traffic class in the next subcycle; i.e., GATE message is delayed in comparison with the standard definition, as shown in Figs. 4.15 and 4.22. This delay enables the OLT to obtain the more up-to-date buffer occupancy information from all ONUs and to increase the system efficiency.

However, the presented model does not fully address the QoS issue in EPON because it does not take into account the QoS requirements of the lowest priority traffic. Consequently, the transmission of the AF traffic could utilize the entire bandwidth, and hence, the delay of the BE traffic could be practically infinite. In our simulation model, called the hybrid granting protocol with priority-based scheduling HG(PBS), we further develop the HG model with the QoS support for the lower-priority traffic [48, 51–52]. We implement and analyze the Intra-ONU scheduling algorithms for the AF and BE traffic classes in order to achieve fairness and bandwidth allocation based on packet priorities.

A. *Inter-ONU scheduling (DBA algorithm)*

The AF and BE traffic classes have bursty transmission characteristics, and therefore, bandwidth allocation for this classes is based on the DBA with QoS model [55], i.e., the limited IPACT scheme presented in [42, 47]. The limited IPACT scheme grants the requested number of bytes to the ONU, but not more than the predefined value called the minimum guaranteed bandwidth. However, due to the bursty nature of Ethernet traffic, some ONUs might have less traffic to transmit, while other ONUs require more than the minimum guaranteed bandwidth. To improve the limited bandwidth allocation algorithm, authors of [55] exploit

Fig. 4.22 OLT information latency

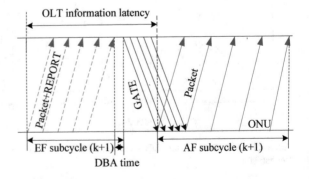

excessive bandwidth (saved from the under-loaded ONUs) by fairly distributing it among the highly loaded ONUs. In the presented scheme, called the weighted inter-ONU DBA algorithm, all ONUs could be partitioned into two groups—the under-loaded and the overloaded ONUs, based on their requested bandwidth. ONUs that request bandwidth less than the minimum guaranteed are in the first group, while the ONUs that require more bandwidth than the minimum guaranteed belong to an overloaded group. In the proposed solution, the total bandwidth saved from the under-loaded ONUs is reallocated to an overloaded group in order to improve system efficiency.

Now, a REPORT message is sent at the end of the transmission window for each ONU. In such a way, the OLT receives the information about buffer occupancy which is updated up to the moment which is one cycle (one EF and one AF/BE subcycle) ahead the starting time of the next AF/BE subcycle. Now, the OLT can allocate more bandwidth to the next AF/BE subcycle in case the MTCT limitation is not exceeded.

The total available bandwidth in the DBA-CL can be expressed as follows:

$$B^{\text{total}} = R \times \left(T_{\text{cycle}}^{\max} - 2 \times T_{\text{g}} \right) \tag{4.16}$$

where R is OLT link capacity, T_{g} is guard interval, and is the MTCT. The granted EF subcycle window size for ONU$_i$ can be calculated as follows:

$$B_{i,k+1}^{\text{EF}} = R \times \left(t_{i,k+1}^{\text{start}} - t_{i,k}^{\text{start}} \right) + L_{\text{report}}, \tag{4.17}$$

where $t_{i,k}^{\text{start}}$ is EF subcycle transmission start time of ONU$_i$ in kth DBA-CL, and L_{report} is bit length of the REPORT message. The transmission of AF and BE traffic is based on the weighted scheme. According to this, the amount of AF and BE traffic in kth DBA-CL in ONU$_i$, i.e., the granted AF/BE subcycle window size, can be calculated as follows:

$$B_{i,k}^{\text{AF}} = \begin{cases} a_{i,k}, B_k^{\text{over}} \le B_k^{\text{under}} & \text{or} \quad a_{i,k}, B_i^{\min} \le B_{i,k}^{\text{EF}} \\ B_i^{\min} - B_{i,k}^{\text{EF}} + B_k^{\text{under}}, & \text{otherwise} \end{cases}, \tag{4.18}$$

$$B_{i,k}^{\text{BF}} = \begin{cases} b_{i,k}, B_k^{\text{over}} \le B_k^{\text{under}} & \text{or} \quad b_{i,k} \le B_i^{\min} - B_{i,k}^{\text{EF}} - B_{i,k}^{\text{AF}} \\ B_i^{\min} - B_{i,k}^{\text{EF}} - B_{i,k}^{\text{AF}} + B_k^{\text{under}}, & \text{otherwise} \end{cases}, \tag{4.19}$$

$$B_k^{\text{under}} = \sum_{i=1}^{M} \left(B_i^{\min} - B_{i,k}^{\text{EF}} - a_{i,k} - b_{i,k} \right), \tag{4.20}$$

$$B_k^{\text{over}} = \sum_{i=1}^{K} \left(B_{i,k}^{\text{EF}} + a_{i,k} + b_{i,k} - B_i^{\min} \right), \tag{4.21}$$

where

$a_{i,k}$ is the requested AF subcycle window size by ONU$_i$,

$B_{i,k}^{AF}$ is the granted AF subcycle window size for ONU$_i$, where $B_{i,k}^{AF} \leq a_{i,k}$,

$b_{i,k}$ is the requested BE subcycle window size by ONU$_i$,

$B_{i,k}^{BE}$ is the granted BE subcycle window size for ONU$_i$, where $B_{i,k}^{BE} \leq b_{i,k}$,

M is the number of under-loaded ONUs, and K is the number of overloaded ONUs,

K is the number of overloaded ONUs,

B_k^{under} is the total excessive bandwidth saved by M under-loaded ONUs,

B_k^{over} is the total excessive bandwidth requested by K overloaded ONUs

B. *Intra-ONU scheduling* scheduling

As discussed in the Sect. 4.5.2, the basic scheduler defined in EPON is a strict priority scheduler (SPS). This scheduler always favors the transmission of the highest priority traffic which can cause indefinite delays and decrease the levels of performance of lower-priority traffic [56–57]. As illustrated in Fig. 4.15, more packets arrive in the buffer during the waiting interval in ONU. Highest priority traffic that has arrived during the waiting time is scheduled for transmission before lower-priority traffic that has arrived before the ONU generated the REPORT message. Although the SPS mechanism protects the transmission of highest priority traffic in terms of delay and jitter, the lower-priority traffic has to wait for transmission and its delay can be indefinitely stretched for several cycles. In the HG protocol, the amount of bandwidth for EF traffic is determined one cycle ahead and the MTCT does not allow high priority traffic to completely utilize the entire bandwidth. However, the protocol defines one subcycle for the transmission of the AF and BE classes where the AF traffic is always prioritized over the BE traffic class. Hence, the transport of the lowest priority traffic has to be solved, because in this algorithm, the AF traffic is always transmitted before the BE traffic class. In that way, AF traffic can occupy the whole remaining bandwidth and BE traffic transport will stretched for several cycles. Even though BE traffic is not delay-sensitive, this unlimited delay could seriously degrade the performance of the entire network.

To overcome the above-stated problems, we suggest the implementation of the scheduling mechanism based on packet priorities, i.e., the PBS (priority-based scheduling) mechanism. In the PBS mechanism, the packets that have arrived before the waiting interval are, based on their priorities, first scheduled for transmission. If the window size can accommodate more packets for transmission, the algorithm will schedule the packets that have arrived during the waiting time on the basis of their priority. In this way, the scheduling mechanism does not allow the higher-priority traffic to monopolize the use of bandwidth.

Now, in the HG protocol, which is essentially concerned with the transmission of EF traffic only, we implement the PBS scheme for the transmission of both AF and BE traffic. Now, AF traffic is not absolutely prioritized and BE traffic has the opportunity to be transmitted without the 'unlimited' delay. The HG scheduler ensures that higher-priority traffic is served with less average delay and jitter and does not allow the EF traffic to occupy the entire bandwidth. Intra-ONU scheduler at each ONU guarantees minimal bandwidth allocation to each service class with different priorities for each user. It can also guarantee the minimal bandwidth allocation to each traffic queue. The analyzed solution can fulfill the QoS requirements in EPON and improve the transmission efficiency of multimedia traffic.

Chapter 5
Multichannel EPON

5.1 Introduction

Most of the FTTH deployments in recent years have been based on the industry's standard technologies such as EPON and GPON. The success of these deployments has led to significant innovations in both system architecture and components which are used for building these systems. However, the rapid increase in bandwidth demands driven by media consumption, such as file sharing, high-definition video, gaming, and many others, has imposed the need for the next generation of passive optical networks that will inevitably be far more advanced in comparison with what is typically being deployed today.

Today, we may single out two technologies in the industry which qualify as the likely candidates for the next fiber access technology [65]:

- 10GPON, as a continuation of GPON and/or EPON;
- WDM PON, taking advantage of the wavelength domain.

Each approach has its own advantages and its own issues, even though the progress related to both new technologies has accelerated in recent years. Nevertheless, from a service-level perspective, no other PON technology, including 10GPON, offers the same bit rate to each home than the bit rate the WDM PON can provide. WDM PON offers 1.250 Mbps per-user bandwidth that is comparable only with point-to-point systems, but WDM PON leverages PON fiber plant of a lower cost. At the same time, in a typical GPON or EPON system, all subscribers use the same common wavelengths, i.e., they share the fiber infrastructure in a way explained in the previous chapters. Each subscriber transmits over the same fiber, but the time during which they are allowed to 'occupy' the fiber is allocated by the OLT at the central office through the implementation of various bandwidth allocation algorithms. This concept, where many users share a common fiber, helps minimize the fiber infrastructure required in FTTH deployment but at the same time, it presents the main limitation for the realization of the high-speed access

© Academic Mind and Springer International Publishing AG 2017
M. Radivojević and P. Matavulj, *The Emerging WDM EPON*,
DOI 10.1007/978-3-319-54224-9_5

network. Namely, while the equipment in each home is capable of transmitting at over 1.250 Mbps, it can only transmit data in the allocated time on the fiber, and therefore, it is common for each subscriber in a legacy PON system to only achieve sustained data rates of around 30 Mbps [66].

As described in the previous chapters, an EPON network presents an efficient solution for the implementation of the access network since it is based on the use of low-cost optical components and the popular Ethernet technology. However, with the increasing number of users in the single-channel TDM EPONs, the duration of the cycle becomes significantly larger and the total delay is rapidly increasing. Moreover, the development of new applications and services, primarily multimedia applications, has caused the increase in the need for higher bandwidth and higher speeds in the access network and these cannot be fully realized with the conventional single-channel EPON network. In such circumstances, the implementation of the WDM technology in the EPON network, i.e., the realization of the WDM EPONs, is considered to be the best solution for the implementation of converged triple-play networks [45].

The implementation of WDM technology would allow access network operators to respond to user requests for service upgrades and network evolution. Moreover, the deployment of WDM technology adds a new dimension to current TDM EPONs whereby the benefits of the new wavelength dimension are manifold. Among others, it may be exploited to increase network capacity, improve utilization of optical infrastructure, and improve network scalability by accommodating more end-users and separating services and service providers. Some advantages of the WDM EPONs include:

- Dedicated bandwidth, guaranteed QoS;
- Physical P2MP, logical P2P;
- Protocol and data rate transparency;
- Simple fault localization;
- Better security;
- Possibility for service differentiation;
- A major drawback of WDM PON, in comparison with the TDM PON, such as EPON or GPON, is the requirement to provide multiple optical ports at the CO. It is therefore necessary to use highly integrated multiple channel transmitter and receiver arrays. Optimized array designs also offer the possibility of the reduction of the electrical power consumption and heat dissipation. Moreover, the main challenges that have impacted the future WDM PON deployments, namely the cost and port density, are being recently addressed through the realization of the low-cost integrated components.

Until now, a significant number of different solutions for the implementation of the WDM technology in EPON have been presented in literature but neither of these solutions has established itself as the dominant one. However, the most discussed architecture in literature is the point-to-multipoint broadcast star architecture. There is variety of reasons for this—the implementation of low-cost

Ethernet equipment, the potential upgrade of the existing EPONs, and many others. Hence, most attention is given to the point-to-multipoint WDM EPONs, i.e., to the analysis of the possible upgrade of the existing single-channel TDM EPONs for the realization of the hybrid TDM/WDM EPONs. In this case, the upgrade of a single-channel to WDM EPON should not impose any particular WDM architecture and should allow the incremental upgrade on a per need basis.

In the following sections, we discuss the different options for WDM technology implementation, along with the potential architectures and protocol extensions.

5.2 Network Architecture

The system architecture in a WDM PON network is not significantly different from that of a more traditional EPON or GPON system, although the network operation is entirely different. Nevertheless, virtually all PON technologies rely on some form of WDM to enable bidirectional communications. For example, in a typical GPON system, the upstream communication runs at 1310 nm wavelength, while the downstream traffic runs at 1490 nm. A third wavelength at 1550 nm is used for video overlay. Therefore, the utilization of WDM in PON systems already exists. However, in a typical GPON or EPON system, all subscribers use those same common wavelengths, i.e., one optical trunk link is shared among 16–64 users. In that case, each subscriber operates at the same wavelength and is allocated a bandwidth portion (1/16nd, 1/32nd, 1/64nd timeslot) on the main fiber.

The simplest approach for the implementation of the WDM EPON is to build a WDM PON that employs a separate wavelength channel from the OLT to each ONU, one for the each upstream connection, and one for the each downstream direction, Fig. 5.1. This represents the most common case of WDM PON implementation in literature, and it is referred to as 'wavelength per ONU assignment.' Contrary to more traditional PON architectures, in WDM PON, each home is assigned its own wavelength and has the continual use of the fiber at that wavelength. Moreover, this approach creates a point-to-point link between the OLT and each ONU, in contrast to the point-to-multipoint topology of the regular TDM EPON. The absence of the point-to-multipoint links leads to a system in which each ONU in the system can operate at a rate of up to the full bit rate of a wavelength. Moreover, different wavelengths may operate at different speeds, and, hence, different sets of wavelengths may be used to support different independent EPON sub-networks, where all operate over the same fiber infrastructure. Since point-to-point connections between the OLT and ONUs are realized in the wavelength domain, no point-to-multipoint media access control is required. This greatly simplifies the MAC layer and removes the distance limitation posed by the ranging and the DBA protocols. A very high-level view of a WDM PON network is illustrated in Fig. 5.1.

The passive node (PN) or the remote node (RN) can be made of either a power splitter or a passive wavelength router. Consequently, three basic WDM PON

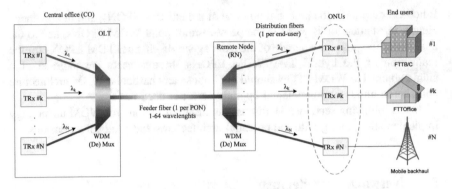

Fig. 5.1 WDM EPON architecture

architectures gain most attention in the literature: broadcast and select WDM PON architecture, Arrayed waveguide grating (AWG)-based wavelength routing PON, and wavelength routing with spectrum slicing PON, Fig. 5.2. Their characteristic, among other, includes:

1. Broadcast and select WDM PON architecture [67]:

   - Passive splitter/combiner in a passive mode;
   - Unique filter in ONU;
   - Individual wavelength upstream;
   - Broadcast security issues;

2. AWG-based wavelength routing PON:

   - Multifrequency laser (MFL);
   - Band splitter (BS);
   - Low insertion loss, 5 dB;
   - Universal Rx;
   - Wavelength specific Tx;
   - Periodic routing behavior;

3. Wavelength routing with spectrum slicing based on colorless ONUs:

   - AWG with identical ONUs;
   - Single shared wavelength upstream (TDMA);
   - Bidirectional OLT using a circulator;
   - Colorless ONU is mandatory and should include:

     - OAM and inventory issue with colorless ONUs;
     - LED or
     - Tunable laser or
     - Injection locked FP lasers or
     - Reflective semiconductor optical amplifiers;

   - Broadband LEDs and spectrum slicing give poor power budget.

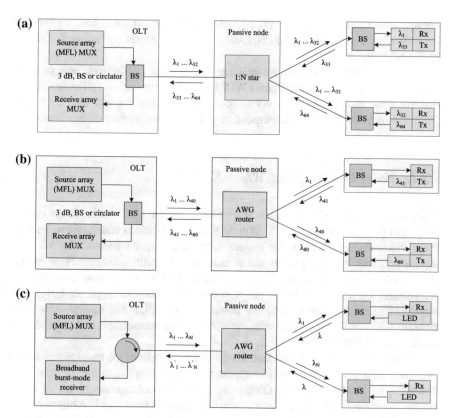

Fig. 5.2 Basic WDM PON architectures: **a** Broadcast and select WDM PON with splitter/combiner in passive node; **b** AWG-based wavelength routing PON; **c** Wavelength routing with spectrum slicing based on LED (*MFL* multi frequency laser, *BS* band splitter, *Tx* transceiver, *Rx* receiver)

A power splitter distributes all incoming signals evenly into all output ports, requiring a wavelength filter at each ONU. Broadcast and select WDM PON architecture based on the implementation of the passive optical splitter is the most inexpensive and simplest solution for the realization of the WDM EPONs. On the other hand, the AWG has been a successful device in WDMindustry. It has been used in many long-distance WDM systems as a multiplexer/demultiplexer. It routes each specific wavelength to a unique output port, separating multiple wavelengths at the same time. Its cyclic wavelength property enables the AWG to be used at the RN, both as a multiplexer and a demultiplexer at the same time. In the simplest case, in the downstream direction of the WDM EPON (Fig. 5.2b), the wavelength channels are routed from the OLT to the ONUs by an AWG router.

In our work so far, we have concentrated on the first architecture as currently dominant approach for the implementation of WDM technology in the access network. Namely, this approach allows the smooth migration from the single-channel TDM EPONs to WDM EPONs and offers the best balance between the implementation costs on the one hand, and system efficiency and performances on the other hand.

In the following sections, we analyze the system architecture and protocol extensions in the point-to-multipoint WDM EPONs with optical splitter/combiner. However, the models with the AWG router implementation will also be briefly discussed in Sect. 5.8 with the purpose of a further clarification of this issue.

5.2.1 Point-to-Multipoint WDM EPON

In the point-to-multipoint broadcast star architecture, the OLT resides in the central office and provides adaptation to different video applications and services, such as VoD, video conferencing, voice transmission, and Internet access, Fig. 5.3. The service contents are multiplexed and broadcasted downstream where these broadcast signals are further transmitted into an individual fiber connected to ONUs by the splitter. On the other hand, the upstream transmissions from all ONUs are controlled and scheduled in a timely manner by the OLT and thus packets from different ONUs will not collide with each other when accessing the fiber trunk that connects the splitter with the OLT.

When it comes to the WDM technology implementation, the transceivers in the OLT and ONUs must be able to support transmission on multiple wavelengths. Accordingly, in the OLT and ONUs, an array of fixed laser/receivers or more tunable laser/receivers have to be implemented. Besides the logical architecture, the election of the wavelength band, i.e., the support for the Coarse or Dense WDM technology, is another issue that must be resolved.

Coarse WDM (CWDM) is a form of wavelength division multiplexing that, as opposed to the Dense WDM (DWDM), has wider spacing between the used wavelengths. ITU specification G.694.2 [68] defines 18 channels in five bands with 20 nm spacing and with a typical capacity ranging from 50 Mb to 2.7 Gb. Both technologies support multiprotocol transport, but in the past, CWDM has showed

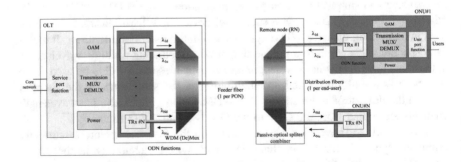

Fig. 5.3 WDM EPON point-to-multipoint architecture

significant market growth due to its low-cost attributes and design simplicity. An example of the CWDM PON is a triple-play PON service where the 1550 nm wavelength channel is used for optional video CATV, the 1490 nm wavelength channel is used for downstream voice and data, while the 1310 nm wavelength channel is used for upstream transmission [1].

In comparison with the DWDM technology, the advantages of the CWDM are:

- Smaller space requirements (30%);
- Support for SMF or MMF cable;
- Can use both LED and laser for power;
- Larger individual payloads per channel;
- Smaller and cheaper wave filters;
- Cost savings.

However, CWDM has less capacity and a shorter range than DWDM and does not support management function. On the other side, DWDM has wavelength spacing that is significantly smaller than that of CWDM, typically less than 1.0 nm, because DWDM has been developed to transmit many wavelengths in a limited spectrum region where an erbium-doped fiber amplifier (EDFA) can be used. It uses a 1550 and 1600 nm bands which have minimum attenuation for long-distance routes, and support from 2 to 256 and more channels. DWDM typically operates at 2.5, 10, and 40 Gb, and with the EDFA implementation, it can be used on long distances. In comparison with CWDM, DWDM is a more complex technology that requires more space, more power, and the implementation of the expensive EDFA amplifiers. However, DWDM PON is expected to be very useful for providing enough bandwidth to many subscribers, and it is regarded as the ultimate PON system for future usage.

The further development of the WDM EPON system demands specification of the transmitter and receiver options. According to the authors of [69], optical sources could be classified into four groups, depending on the way wavelengths are generated: a wavelength-specified source, a multiple-wavelength source, a wavelength-selection-free source, and a shared source. The multiple-wavelength source is applicable only to the OLT, the shared source is applicable to the ONU, and the remaining two are applicable to both. A receiver module consists of a photo-detector (PD) and its accompanying electronics for signal recovery. Common PDs are positive-intrinsic-negative (PIN) and avalanche (APD) photodiode and find different applications according to the required sensitivity. Electronic parts, usually composed of a preamplifier, main amplifier, clock, and data recovery circuits (CDRs), depend on the protocol used on each wavelength. Since each wavelength can work separately in a WDM PON, each receiver can be configured differently. The further discussion of transmitter and receiver options is left to vendors, and, hence, in the next section, we discuss and analyze the logical structure of the OLT and ONUs.

5.2.2 Logical Structure of the OLT

As described above, all ONUs in a WDM EPON network have identical structure. The largest system change, in comparison with other PON architectures, comes at the OLT. The WDM EPON OLT is quite complex compared with single-channel systems. Since each subscriber gets the benefit of a full wavelength to their home, this also requires that each subscriber has their own dedicated transceiver in the OLT as well. Although theoretically a tunable laser or an array of fixed-tuned laser can be implemented in OLT, the actual implementation of a tunable laser would not be an advantage. In case of the implementation of the tunable laser, only a single wavelength can be used at a time and the currently available bandwidth of the single-channel system would not be expanded. Moreover, this solution would provide even less bandwidth since laser tuning has to been taken into account. Hence, the OLT is equipped with multiple fixed transmitters for simultaneous downstream transmissions, and multiple fixed receivers that are constantly receiving data transmitted by ONUs in all upstream channels. Figure 5.3 shows Nth channel WDM EPON in which N upstream and N downstream wavelength channels are utilized. Now, the OLT schedules ONUs for transmission over a defined number of upstream channels.

As shown, the OLT consists of the service port function, ODN interface, OAM module, power module, and transmission mux/demux, Fig. 5.4. ODN interface is responsible for the insertion of packets into downstream channels and the extraction of packets from the ONU transmissions. The service port function includes inserting Internet packets into the backbone network and extracting packets from the backbone connections [70]. The transmission MUX/DEMUX provides connections between the service port function and the ODN interface in which different services are multiplexed into downstream channels.

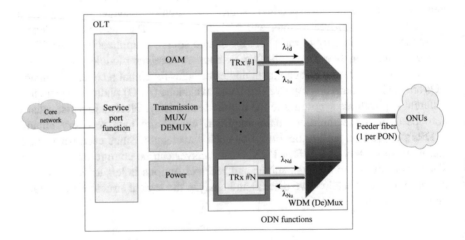

Fig. 5.4 Logical structure of the OLT unit

5.2.3 Logical Structure of the ONU

Even though the ONU could be realized through the implementation of either tunable laser/receivers or fixed laser/receiver arrays, managing ONUs with different WDM capabilities in service provider networks can be extremely difficult so it is very unlikely that providers would implement both solutions. Moreover, in the field, every ONU can be equipped with one or more fixed transmitters, allowing for an incremental upgrade depending on the traffic demand at the ONU.

Contrary to OLT constraints, the WDM capabilities of an ONU do not limit the WDM capabilities of the entire EPON. Therefore, as previously explained, a tunable laser or an array of fixed lasers could be implemented. The OLT is able to schedule a transmission from an ONU on any wavelength supported by that ONU with either laser type. In the downstream direction, the OLT could transmit data to an ONU on any wavelength contained in the set of wavelengths supported by that ONU. This implementation allows the OLT to take advantage of the bandwidth available on other wavelengths. At the same time, ONUs can be upgraded incrementally as needed, by means of adding new fixed and tunable lasers.

With this implementation, an ONU can be scheduled for transmissions simultaneously over a number of wavelength channels. The main functional blocks in the ONU include the transmission ODN interface, OAM module, MUX/DEMUX, power module, and user port function, Fig. 5.5. The ODN interface in the WDM system is different from the single-channel system since it comprises multiple fixed transmitters and receivers. Each transmitter is configured to operate at one of the upstream wavelengths. Each downstream channel is attached to a receiver for the constant reception of broadcast signals from the OLT. The MUX multiplexes customer interfaces to the ODN interface where the demultiplexed downstream flow is forwarded to the corresponding users. The MUX structure includes the logical queue structure for the transmission of frames. This function can be extended to support differential services function [70]. In that case, the queue structure is replaced by an array of priority queues. When the end user packet arrives, it is placed into one of the priority queues according to the defined QoS mechanisms. The user port function provides the necessary interface processes to individual end users connected to the ONU and allows them access to the network system.

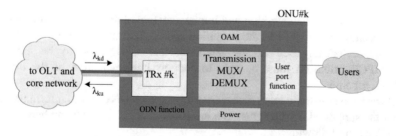

Fig. 5.5 Logical structure of the ONU units

5.3   Information Exchange

In order to define communication in WDM EPON, the information exchange between the OLT and ONUs should not impose a practical WDM architecture on the ONUs. Moreover, with the defined information exchange, the OLT should be able to schedule transmission to and reception from ONUs on any wavelength supported by that ONU. Accordingly, the OLT must know the ONU architecture, i.e., if the ONU supports fixed or tunable lasers. Further, the OLT must know which wavelengths are supported by each ONU for transmission and reception.

The most efficient way for one ONU to inform the OLT about the supported wavelength range is to send this information during the discovery and registration process defined by the MPCP. After that, the OLT chooses one wavelength and informs the ONU that the chosen wavelength will be used for transmission and data reception. With this mechanism, the media access control mechanism is expanded to two dimensions—time and wavelength. It means that the allocation algorithm in the OLT is no longer defined as DBA. By introducing transmission on various wavelengths, allocation algorithm is now defined as the dynamic wavelength and bandwidth allocation (DWBA). The DWBA algorithm now manages not only the bandwidth allocation, but the wavelength allocation as well. However, the OLT is unaware of the ONUs-supported wavelengths before the discovery and registration process.

The first solution to that problem is to configure each ONU to provide discovery services on all supported wavelengths. The second solution is to provide control services only on a single-wavelength channel that is solely defined for registration and discovery. This wavelength could be either the original wavelength that was defined for EPON usage or some other predefined wavelength. For ONUs with tunable laser/receivers, this approach requires that they have a tunable receiver range that covers the discovery and registration wavelength and that the receiver is initially tuned to this wavelength. For ONUs with fixed laser/receiver arrays, this approach requires that the discovery and registration wavelength is covered by the array. This final decision is left to the implementers.

5.4   MPCP Extensions

In order to support the definition and implementation of an additional dimension in the allocation algorithm, i.e., the definition and implementation of the DWBA, the control mechanism used in the single-channel EPON has to be extended. The MPCP extension must be able to support implementation of the different DWBA scheme.

During the registration and discovery process, each ONU has to inform the OLT about the supported wavelengths. The information about the ONU architecture has to include the following:

- Optical transmitter and receiver type (tunable or fixed);
- In case of tunable transmitter/receiver, the tuning time as well;
- List of supported wavelengths for both the transmitter and the receiver.

This information allows the OLT to know the following:

- Which wavelengths can be used when communicating with this ONU;
- Whether the wavelengths can be used concurrently or not (fixed array or tunable);
- In case it consists of a tunable transmitter and/or receiver, the tuning time required to change the wavelength channel used for transmission or reception by an ONU.

The authors in [71] propose the following extensions to the MPCP discovery/registration process. The MPCPDU used in the ONU registration process has to be extended with the following fields:

- Transmitter_type field (Table 5.1);
- Transmitter_tuning_ time field: an integer value with a timescale in microseconds (16 bits required);
- Transmitter_supported_wavelengths field: a bitmap representing the supported wavelengths;
- Receiver type field (Table 5.2);
- Receiver tuning time field: an integer value with a timescale in microseconds (16 bits required);
- Receiver supported wavelengths field: a bitmap representing the supported wavelengths.

The fields 'Transmitter_supported_wavelengths' and 'Receiver_supported_ wavelengths' require the definition of wavelength channels in a standard well-known numeric fashion. In the first place, in order to define bitmap representing the supported wavelengths, the authors define the parameter called 'wavelength identifier.' The wavelength identifier parameter is defined through a standard set of wavelengths and a unique identifier for each wavelength.

Table 5.1 'Transmitter type' field definition

Value	Transmitter type	Number of bits
0	No WDM	2
1	Fixed	2
2	Tunable	2
3	For both	2

Table 5.2 'Receiver type' field definition

Value	Transmitter type	Number of bits
0	No WDM	2
1	Fixed	2
2	Tunable	2
3	For both	2

The supported wavelengths could be defined through the two-level hierarchical bitmap (waveband number and wavelengths bitmap), or a flat bitmap with a single bit for each of the 176 possible wavelengths. The actual wavelength band can be defined according to the ITU-T standard for DWDM channel spacing in ITU-T G.694.1 [72], which specifies channel spacing of 12.5, 25, 50, or 100 GHz anchored to 193.1 THz or 1552.524 nm. For the reasons of simplicity, the authors suggest the support for the C-band in the first step of WDM implementation with a channel spacing of 25 GHz, i.e., 176 channels. Actually, the proposed channel spacing of 25 GHz allows the coverage of the entire C-band with 175–177 channels, but authors define support for 176 channels since this number is divisible by 8. This implementation would provide up to 1.75 Tbps of bandwidth on a single strand of i.e., 191.55–195.925 THz or 1565.08–1530.14 nm wavelength grids.

In the two-level hierarchical bitmap (waveband method), the two fields represents the supported wavelength. The first field ('Rx_waveband/Tx_waveband') specifies the waveband for the supported wavelengths. The second field specifies a supported wavelengths bitmap that specifies which wavelengths within the waveband are supported. The only restriction would be that the wavelengths supported in an ONU would be restricted to 16 (two bytes field) and would have reside in the same waveband. In case of the C-band support, the waveband would be expressed as an integer that represents steps in wavelengths of 16×25 GHz channels. Each waveband is 16×25 GHz or 400 GHz wide. Since the method has to support 176 channels, the waveband is expressed as a number from 0 to 11, so that along with 16 wavelengths in the 16-bit wavelength bitmap, the total of 176 wavelengths is supported. Moreover, the waveband could easily be expanded to support up to 256 wavebands.

The second method, flat bitmap method, is the method where we have 176 wavelength identifications, one for each 25 GHz channel in the C-band and then we use a 176-bit (or 22 byte) bitmap to express which of the wavelengths in the C-band are supported. This removes the restrictions which refer to fact that an ONU can support up to 16 wavelength channels and that the wavelength channels are contained in the same 400 GHz waveband at the expense of a much larger bitmap for expressing supported wavelengths [71].

5.4.1   Wavelength Band Announcement

As we have previously explained, the best and the most efficient approach for one ONU to report its supported wavelength band is during the registration and recovery process since after that no changes in the supported wavelength band will be taken into account. Accordingly, the REGISTER_REQ MPCP message has to be modified in order to support wavelength band announcement.

Figures 5.6 and 5.7 show the structure of the REGISTER_REQ MPCPDU with the waveband wavelength identification (ID) method and the flat bitmap method.

Octets	Field
6	Destination address
6	Source length
2	Length/Type=88-08
2	Opcode=00-04
4	Timestamp
1	**Flags**
1	Pending Grants
2	**Rx_tunning_time**
2	**Tx_tunning_time**
1	**Rx_waveband**
2	**Rx_supported_wavelenths**
1	**Tx_waveband**
2	**Tx_supported_wavelenths**
28	Pad/Reserved
4	FCS

Bit	Field
0	**Wavelength_id_type=0**
1	register
2	reserved
3	deregister
4-5	**Rx_type**
6-7	**Tx_type**

Fig. 5.6 REGISTER_REQ MPCPDU with waveband wavelength ID method

The 'Flags' field is used to specify wavelength ID method and the transmitter's type. Namely, bit 0 specifies which of the wavelength ID methods is used, Fig. 5.6. This approach allows the support of both of the suggested methods. The transmitter type and receiver type fields use four reserved bits in the Flags field, specifically bits 4 and 5 for receiver type and bits 6 and 7 for transmitter type. The 'Rx_tuning_time' and 'Tx_tuning_time' fields use 2 byte field with a time scale of microsecond. For the wavelength ID method, authors suggest the use of two fields for the specification of upstream/downstream communication, namely the 'Rx/Tx_waveband' and 'Rx/Tx_supported_wavelenghts.' Furthermore, the authors recommend one byte field for the mapping of the waveband method. The 16-bit bitmap for the wavelengths is defined and placed immediately after the byte containing the waveband field, Fig. 5.6.

In order to specify flat bitmap method, a 176-bit (22 byte) bitmap for both receive and transmit supported wavelengths have to be specified. However, there is not enough space in the REGISTER_REQ message for this, and therefore, the

Octets	Field
6	Destination address
6	Source length
2	Length/Type=88-08
2	Opcode=00-04
4	Timestamp
1	**Flags**
1	Pending Grants
2	**Rx_tunning_time**
2	**Tx_tunning_time**
16	**Rx_supported_wavelenths**
16	**Tx_supported_wavelenths**
2	Pad/Reserved
4	FCS

Bit	Field
0	**Wavelength_id_type=1**
1	register
2	reserved
3	deregister
4-5	**Rx_type**
6-7	**Tx_type**

Fig. 5.7 REGISTER_REQ MPCPDU with flat bitmap wavelength ID method

authors propose a 128-bit bitmap for both receive and transmit supported wavelengths (immediately after the tuning time fields). These 128 bits represent the first 128 wavelengths of the entire 176 wavelengths from the C-band. Figure 5.7 shows the REGISTER_REQ MPCPDU for this method.

5.4.2   Upstream Coordination

In the upstream direction, the OLT has to assign a specific wavelength to each ONU through the additional fields in the GATE message. This could be achieved through the definition of the specific wavelength field that would be used along with every transmission grant issued by the OLT. Figure 5.8 shows the two possible extensions of the GATE MPCPDU in case of the waveband wavelength ID method implementation (Fig. 5.8a) and in case of the flat bitmap wavelength ID method implementation (Fig. 5.8b).

(a)

Octets	Field
6	Destination address
6	Source length
2	Length/Type=88-08
2	Opcode=00-02
4	Timestamp
1	Number of Grants/Flags
4	Grant #1 Start Time
2	Grant #1 Length
4	Grant #2 Start Time
2	Grant #2 Length
4	Grant #3 Start Time
2	Grant #3 Length
4	Grant #4 Start Time
2	Grant #4 Length
4	Synch time
1	**Grant #1 Channel (Wavelength)**
1	**Grant #2 Channel (Wavelength)**
1	**Grant #3 Channel (Wavelength)**
1	**Grant #4 Channel (Wavelength)**
9-39	Pad/Reserved
4	FCS

(b)

Octets	Field
6	Destination address
6	Source length
2	Length/Type=88-08
2	Opcode=00-02
4	Timestamp
1	Number of Grants/Flags
4	Grant #1 Start Time
2	Grant #1 Length
1	**Grant #1 Channel (Wavelength)**
4	Grant #2 Start Time
2	Grant #2 Length
1	**Grant #2 Channel (Wavelength)**
4	Grant #3 Start Time
2	Grant #3 Length
1	**Grant #3 Channel (Wavelength)**
4	Grant #4 Start Time
2	Grant #4 Length
1	**Grant #4 Channel (Wavelength)**
2	Sync time
9-39	Pad/Reserved
4	FCS

Fig. 5.8 GATE MPCPDU: **a** waveband wavelength ID method; **b** flat bitmap wavelength ID method

5.4.3   Downstream Coordination

In the downstream direction, the OLT has to assign a specific wavelength for the
transmission from the OLT to an ONU, and hence a new MPCPDU with the
information about the wavelength supported by an ONU has to be defined. This
new MPCPDU specifies changes in the receiving wavelength for an ONU during
the OLT—ONU communication. The current message in the communication is
received on the previously used wavelength (the ONU must acknowledge this
message) and from that point all communication is realized on the new wavelength,
until another wavelength switch is initiated. Actually, the described procedure
requires the definition of the two control messages RX_CONFIG and
RX_CONFIG_ACK, Figs. 5.9 and 5.10. Both control messages have 'Flags' fields
in which byte 0 defines the wavelength identification method. Value 0 (byte 0 = 0)
specifies the hierarchical bitmap while value 1 (byte 0 = 1) specifies a flat
bitmap. Besides the 'Flags' field, the RX_CONFIG frame has a
'Reception_wavelengths' field that would either be a 16-bit bitmap or a 128-bit
bitmap depending on which wavelength ID method is used in the supported
wavelength field. The RX_CONFIG_ACK frame consists of a reserved Flags field.

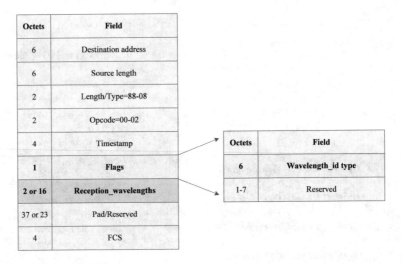

Fig. 5.9 RX_CONFIG MPCPDU

Octets	Field
6	Destination address
6	Source length
2	Length/Type=88-08
2	Opcode=00-02
4	Timestamp
1	Flags
39	Pad/Reserved
4	FCS

Fig. 5.10 RX_CONFIG_ACK MPCPDU

5.5 WDM EPON Resource Allocation

In the single-channel EPONs, control MPCP messages cannot arrive concurrently since they are transmitted by one fiber link. Accordingly, an instantaneous scheduling of the REQUEST_message at the next available upstream time (round-robin policy) is massively deployed. On the other side, in the multichannel EPONs, the OLT has to manage not only the bandwidth in the time domain, but it also has to manage it in the wavelength domain as well. Consequently, in the WDM EPON, the efficient scheduling grants across multiple wavelengths are necessary to fully utilize the network resources. Ideal scheduling algorithm would include a complex optimization function that ensures fair bandwidth allocation with as few wavelength assignment changes as possible [71]. However, this algorithm is too robust and cannot be implemented. Namely, the time required for the implementation of this kind of algorithm is too long in comparison with one polling cycle and the fact that scheduling algorithm has to be executed in every polling cycle, which can be in the order of milliseconds.

Consequently, the bandwidth management for a WDM EPON can be broken into two sub-problems: grant sizing and grant scheduling. Grant sizing determines the size of a grant to an ONU, and grant scheduling determines when and on which wavelength channel the grant should be scheduled. Furthermore, grant scheduling could be realized as follows:

- Separated time and wavelength assignment;
- Joint time and wavelength assignment (i.e., multidimensional scheduling).

5.6   Separate Time and Wavelength Assignment

In case of the separate allocation of bandwidth and wavelength, bandwidth allocation can be managed in the time domain and any of the DBA algorithms proposed so far could be implemented [73]. For wavelength allocation, several wavelength assignment heuristics approaches could be used for wavelength allocation in the upstream as well as in the downstream transmission between the OLT and the ONU. These heuristics have been adapted from the literature focusing on wavelength-routed networks but, contrary to light-path networks, EPONs do not require the implementation of wavelength routing since the links are of a single hop type (ONU to OLT and OLT to ONU). Hence, the presented approaches do not consider the assignment of multiple light-paths between ONUs and the OLT (for ONUs with fixed laser/receiver arrays). Here, we review several heuristic approaches that could be used in case of the static, i.e., dynamic wavelength assignment in WDM EPON.

5.6.1   Static Wavelength Allocation

As stated in this approach, once the wavelength is assigned for the upstream and downstream transmission it remains static. The upstream and downstream transmission can be realized by the use of the same wavelength, but this is not necessarily always so. As simple as it may seem, this approach does not take into consideration any changes in network conditions and does not support any adaptations to the current network status or to the instantaneous bandwidth requirements.

The wavelength could be allocated in the following ways:

- Random Allocation—one of the ONU supported wavelengths is randomly selected and used for the upstream/downstream transmissions;
- Least Assigned Allocation—the wavelength that has the minimum load is allocated to the minimum number of other ONUs in the system;
- Least Loaded Allocation—selects the wavelength that is supported by an ONU that has the least load assigned to it. The load of each ONU is available during the discovery and registration process.

5.6.2   Dynamic Wavelength Allocation

In this approach, wavelength allocation is periodically checked and adjusted, if necessary, to follow eventual changes in the EPON network conditions. The heuristic approaches used in this case require more logic, but are much more efficient in comparison with the previously described static assignment. In the ideal case, wavelength allocation would be synchronized with the bandwidth requirements, i.e., the OLT would monitor the utilization of each wavelength and consequently use this information along with a used heuristic to decide whether it is necessary to change the wavelength or not.

The authors of the [71] suggest two possible heuristics in this paradigm

- Load Shifting—in this approach, the OLT monitors the utilization of wavelengths in each of the ONUs, and when the wavelength load exceeds a certain predefined threshold, the OLT attempt to move the ONU assigned to this wavelength to another wavelength. For the reallocation, the OLT chooses one of the supported wavelengths that have the lowest current utilization. After the reallocation, the algorithm moves to the next heavily loaded wavelength;
- QoS Load Shifting—according to authors, this method would adjust the loading similarly to Load Shifting but would use a QoS measure instead of actual load.

However, even though the dynamic wavelength allocation takes wavelength utilization into account, it is not able to efficiently coordinate the wavelength allocation and bandwidth management. When large time scales are used for wavelength assignment changes, the assignment changes will not allow the bandwidth management to adapt to traffic load changes quickly enough. Since traffic in the access network has bursty transmission characteristics, the performance of the dynamic wavelength assignment will not provide significant improvement in comparison with the static wavelength assignment. Moreover, as we consider the time scale of assignment changes, especially an assignment change for each grant, the dynamic wavelength assignment becomes a joint time and wavelength scheduling problem that requires the implementation of different approaches in order to be successfully resolved.

5.7 Joint Time and Wavelength Allocation

As we have explained and discussed in the previous chapter, the resource allocation in the single-channel EPONs is defined as a bandwidth management problem, and it is concerned with the scheduling of the upstream transmissions, i.e., bandwidth allocation on the single-wavelength channel. However, in WDM EPONs, the bandwidth management problem is expanded to scheduling the upstream transmissions on different upstream wavelengths supported by the ONUs [74]. Hence, the resource allocation can be defined as grant scheduling (wavelength allocation) and grant sizing (bandwidth allocation). Now, the OLT has to decide about not only when and for how long to grant an upstream transmission to an ONU but also about on which of the supported wavelength channels to grant the upstream transmission in order to make the efficient use of the transmission resources. This joint time and wavelength allocation is also called the multidimensional scheduling in the literature.

The first and the mandatory step in the scheduling algorithm implementation is the definition of the scheduling cycle. The scheduler has to increase resource (i.e., channel) utilization and to decrease queuing delays experienced by frames in transit across the EPON. According to the authors of the [74], the cycle length can be defined as GATE-to-GATE delay (GTG), i.e., as the time between two grants allocated to an ONU. As shown in Fig. 5.11, the GTR (GATE-to-REPORT) delay is equal to the transmission time of the grant since the REPORT message is transmitted at the end of the transmission window. The RTG (REPORT-to-GATE) delay includes the propagation delay from an ONU to the OLT [74].

Accordingly, the cycle length can be expressed as follows:

$$T_{\text{GTG}} = T_{\text{GTR}} + T_{\text{RTG}}. \tag{5.1}$$

The GTR delay is completely dependent on the grant size, and the RTG delay is dependent on the implemented scheduling mechanism. The STG (schedule-to-GATE) represents the scheduling delay and is defined as the time between the OLT scheduling an ONU's next grant and the time the grant starts. The STG, along with the grant time, represents the completion time of the ONU's transmission from the point in time it is scheduled (STG+GTR). The minimization of this interval automatically minimizes the STG interval.

The OLT to ONU propagation of the GATE message can be compensated through the inter-leaving with an online scheduler. The RTS (REPORT-to-Schedule) delay is the time interval between the moments in which the OLT receives a REPORT from an ONU and the moment in which the ONU's REPORT is considered for scheduling by the OLT. Thus, REPORT-to-GATE delay is composed of the RTS, GTR and STG delays, hence the cycle length can be expressed as:

$$T_{\text{GTG}} = T_{\text{GTR}} + T_{\text{RTS}} + T_{\text{STG}}. \tag{5.2}$$

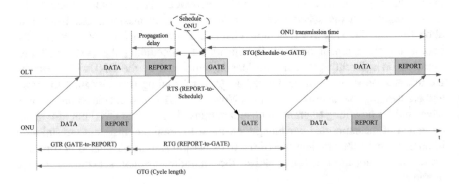

Fig. 5.11 Definition of the scheduling cycle

This analysis and the definitions given above will be used in the following analyses of the two broad paradigms for scheduling grants for upstream transmissions on different upstream wavelengths in WDM EPON—the offline scheduling and online scheduling.

5.7.1 Offline Scheduling

In an offline scheduler, the ONUs are scheduled for transmission when the OLT has received current MPCP REPORT messages from all ONUs. Hence, the REPORT messages from all ONUs, which are usually appended to the end of the data stream of a gated transmission window from the previous cycle, must be received. This mechanism allows the OLT to take into consideration the current bandwidth requirements of all ONUs and to use them for the scheduling and grant sizing, i.e., the scheduling pool contains all ONUs. The scheduling policy is executed after the OLT has received the end of the last ONU's gated transmission window. In accordance with the analyses presented in the previous section, the RTS delay of the last scheduled ONU will be negligible. However, the RTS delay for other ONUs is not negligible and will add additional queuing delays in the ONUs since it introduces the additional delay in the cycle length (GTG) for an ONU, Fig. 5.11.

Offline scheduling mechanism is efficient in terms of bandwidth allocation since the OLT takes the current bandwidth requirements of all ONUs into consideration in the grant sizing and scheduling, but at the same time, waiting for all ONU REPORT messages to be received results in wasted channel capacity. Moreover, the wasted capacity increases as the number of channels increases.

5.7.2 Online Scheduling

The online scheduling mechanism tends to resolve the previously described problems of the offline approach since in the online scheduler the scheduling decisions are made without the complete knowledge of all of the ONUs requirements. In EPONs, MPCP REPORTs arrive over time and the processing time of grant size is known in full once the bandwidth management or DBA algorithm in the OLT computes the grant size from the MPCP REPORT message. Even though MPCP REPORTs arrive over time, the OLT of a single-channel EPON can receive only a single REPORT message at a time because of the single-wavelength channel used for the reception. However, an OLT in the multichannel EPON can receive multiple REPORT messages within a short time interval. Each REPORT message is received on a separate wavelength channel. This framework avoids channel idleness, but makes scheduling decisions based on the REPORT information of a single ONU only and this could result in lower system efficiency.

Until now, two dominant approaches have been discussed in the relevant literature regarding the online scheduling implementation, namely the immediate and the just-in-time online scheduling mechanisms.

A. *Immediate online scheduling*

The online scheduler schedules ONUs one at a time, without considering the bandwidth requirements of other ONUs. For this type of online scheduler, a given ONU is scheduled for upstream transmission as soon as the OLT receives the REPORT message from the ONU. A basic online scheduling policy for the WDM EPON is to schedule the upstream transmission for an ONU on the wavelength channel that is available the earliest among the channels supported by the ONU, which is referred to as the next available supported channel (NASC) policy [74].

Each ONU reports its queue occupancy in the REPORT message, which is appended to the current upstream transmission (Fig. 5.11). Upon the receipt of a REPORT message, the OLT immediately schedules, on the earliest available supported wavelength, the next upstream transmission for the corresponding ONU and sends a GATE delay is composed of the RTS indicating the wavelength and length (in bytes) of the granted transmission to the ONU. As previously explained, in an offline scheduler, one ONU has to wait until the OLT receives REPORT messages from all other ONUs. Moreover, the RTS delay noted for offline scheduling does not exist in an online scheduler since an ONU is scheduled as soon as its REPORT is received, Fig. 5.11. Accordingly, the RTG delay is reduced to the STG delay, i.e., the ONU transmission completion time.

B. *Just-in-Time online scheduling*

Besides the previously described online scheduling, where the OLT schedules grants as soon as a REPORT message is received from an ONU, and offline scheduling, where the OLT waits for REPORT messages from all ONUs before scheduling grants, there is a extremely large range of possible scheduling frameworks. One of them is just-in-time (JIT) scheduling framework that is work-conserving while increasing scheduling control by allowing the channel availability to influence the scheduling process [75]. The name indicates that scheduling is performed in a just-in-time fashion. In online JIT, multiple ONU REPORTs can be considered together when making scheduling decisions, which ultimately results in lower average queuing delay under certain conditions.

Within this scheduling framework, the ONUs are added to a scheduling pool as their MPCP REPORT messages are received by the OLT. When some wavelengths become available, the ONUs in the pool are scheduled together according to the selected scheduling policy across all available wavelengths. Hence, the JIT online scheduling framework avoids channel idleness but allows for the possibility of considering multiple ONU REPORTs during the decision-making process at the OLT. The ONUs that are scheduled in such a way so as to have their transmissions occur shortly after the time they are scheduled are classified as 'imminent' and the

current schedule for these ONUs is considered to be firm. The OLT transmits
GATE messages to these ONUs to inform them of their granted transmission
window. The remaining ONUs are classified as 'tentative' and can potentially
remain in the scheduling pool for the next scheduling round. Alternatively, all
ONUs (i.e., both imminent and tentative) can always be firmly scheduled.

In order to implement this approach within an EPON, the GATE message has to
be transmitted by the OLT at least one RTT before the ONU intends to begin
transmission. Since the ONU has to transmit as soon as the next wavelength
becomes free, the ONU has to be scheduled in the pool at least an RTT before the
next wavelength free time. In other words, the GATE message must be transmitted
soon enough to accommodate the RTT. The most efficient way is to use the largest
RTT in the EPON for the timing of the GATE message transmissions. This
approach ensures that every ONU will receive the GATE message in time.

5.8   Review of the Related Work

In order to solve the scalability issue of classic PONs, one of the earliest
WDM PON architecture proposals was based on the TDMA and AWG concept.
The presented architecture has been referred to as the composite PON (CPON), and
it employs separate fibers for upstream and downstream [76]. This architecture also
introduces an integrated type of device that performs WDM routing in one wavelength window on a single fiber and wavelength independent power combining in
the second wavelength window on another single fiber, Fig. 5.12. However,
although CPON avoids the drawbacks of upstream WDM, it remains limited by the
fact that single frequency laser at the ONU may be economically prohibitive.

The LARNET (Local Access Router NETwork) architecture presented in [77]
attempts to overcome the limitations of the CPON architecture by employing a

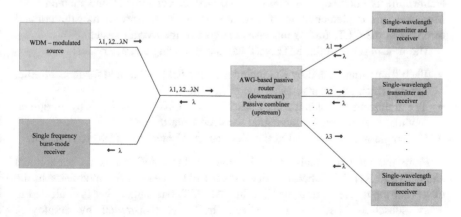

Fig. 5.12 CPON architecture

broad-spectrum source at the ONU, such as an inexpensive LED whose spectrum is sliced by the AWG-based router into different optical bands in the upstream direction. The LED emits a broadband signal, the spectrum which is split into different wavelengths by the AWG, with a loss factor of at least 1/N (for N ONUs). In contrast to CPON, authors further suggest the implementation of a broadband receiver (unlike a single-wavelength burst-mode used in CPON) at the headend, in order to overcome the limitations imposed by the use of a single receiver at the OLT. In this way, the OLT at the CO can receive any wavelength from any of the ONUs. A LED may also be used at the CO, in which case the signal emitted downstream is effectively broadcast to all ONUs. It is possible to have two transmitters at the CO: a 1300 nm LED, for example, broadcasting the signal to all ONUs, and a 1550 nm laser, transmitting only for selected ONUs.

However, the introduction of the low-cost LED at the ONUs, despite lowering the cost of the system, may lead to high power loss, and therefore, the distance from the OLT to the ONU may be considerably reduced in LARNET.

The RITENET architecture avoids the transmitter at the ONU by means of modulating the downstream signal from the OLT and sending it back in the upstream direction [78]. Further, the signal from the OLT is shared for downstream and upstream transmission through the time-sharing mechanisms, i.e., a time frame is split into two parts, one for the downstream, and one for the upstream transmission. Since the same optical channel is used for both the upstream and downstream transmissions, they must be separated on two different fibers. In the proposed architecture, an AWG-based router is used to route the wavelengths. In the OLT, a tunable laser, whose modes match those of the AWG, must be used since both the upstream and the downstream channels have to be shared by the ONUs using TDMA or DBA. In order to avoid channel sharing, the architectures presented in [79] employ an array of transmitters and receivers at the OLT. Although RITENET architecture has some advantages in comparison with CPON and LARNET, such as the availability of symmetrical bandwidth in the downstream and upstream directions, it is more expensive. The number of fibers employed in the architecture is doubled, thus doubling the cost of deployment and maintenance. Moreover, the implementation of a tunable laser or an array of transmitters and receivers at the OLT, further increases the cost of the overall system.

The transmission in the RITENET has the following characteristics:

- Each frame transmitted by CO includes a data field and a field for return traffic, during which the laser remains switched ON;
- The architecture avoids having a laser in each ONU, replacing it by an optical modulator which reuses the signal received from the CO;
- If a single receiver is used at the CO, the ONU must use TDM to access it.

From the previous analyses, it is obvious that LARNET is an economical alternative to RITENET because that uses a LED instead of the modulator in the ONUs for the upstream transmission. In LARNET, one single fiber is dedicated to every subscriber, i.e., the downstream traffic is transmitted by employing wavelengths belonging to the 1550 nm optical window, while for upstream traffic the wavelengths in the 1300 nm window are used (CWDM bi-directionality). In RITENET, however, two fibers are dedicated to every subscriber, where each of the fibers is able to carry the traffic in one direction (space division bi-directionality). These architectures seem to suffer from two main limitations:

- Difficulty in scaling the number of ONUs once the network is laid down;
- Limited number of users, because the fabrication technology imposes limitations on the WGR size.

In order to avoid channel sharing and improve system efficiency, the architecture presented in [79] employs an array of transmitters and receivers at the OLT. Authors proposed a novel WDM access network that establishes a data link layer with a virtual single star topology between end users and the center node over a wide area and provides guaranteed full-duplex Gigabit Ethernet access services to each of over 100 users. The center node employs an optical carrier supply module that generates not only the optical carriers for the downstream signals but also those for the upstream signals.

Besides previous solutions, a multistage architecture was proposed in [80]. This architecture, which is referred to as a multistage WDM PON., exploits the periodic routing property of the AWG so that the reuse of a given wavelength for more than one subscriber is possible. This architecture provides scalability in bandwidth, as well as in the number of users, either by employing additional wavelengths at the CO or by cascading multiple stages of AWGs with increasing AWG coarseness at the each stage. The PON structure is composed of a number of stages interconnected by fiber links, where each stage is composed of passive wavelength routing devices. However, the presented architecture involves additional AWGs and a complex architecture and is therefore not cost-effective.

To resolve the distance issue, as well as the bandwidth issue, a system called DWDM Super PON (SPON) is suggested for bandwidth increase by means of providing several wavelengths in both directions [81]. Although the system cost is lower in comparison with the previous solutions, a DWDM SPON has a bandwidth limitation because a wavelength is shared by many ONUs. Consequently, a PON is limited in its transmission distance and the number of nodes it can support because of a limited power budget. For instance, the maximum transmission distance of EPON is 20 km and the maximum number of supported ONU is 64. The DWDM SPON proposal covers a range of over 100 km with a splitting ratio reaching 2000 with the aid of optical amplifiers (OAs) [82]. Optical amplifiers are placed in the long feeder and after the first splitting stage in order to provide enough power budget to the system. The cost benefits stem not only from sharing resources, but also from consolidating switching sites. The DWDM SPON increases the bandwidth by providing several wavelengths in both directions [83]. In this architecture, each power splitter uses two DWDM channels, one for upstream and the other for downstream direction. All upstream wavelengths are amplified by an EDFA preamplifier before being separated into each receiver at the OLT.

The Stanford University aCCESS (SUCCESS) initiative within the Photonics and Networking Research Laboratory encompasses multiple projects within the access networks. In this paper, we discuss only three of these projects.

In the first project, the authors of [84] propose a hybrid architecture called SUCCESS, which provides migration from the current TDM PONs to the WDM PON while maintaining backward compatibility for users on the existing TDM PONs. The proposed architecture is based on a collector ring and several distribution stars connecting the CO and the users.

According to authors, a novel design of the OLT and DWDM PON ONUs considerably minimizes the system cost in relation to the following facts: (1) tunable lasers and receivers at the OLT are shared by all ONUs on the network to reduce the transceiver count and (2) the fast tunable lasers not only generate downstream data traffic but also provide DWDM PON ONUs with optical CW bursts for their upstream data transmission. Moreover, a presented semi-passive configuration of the RNs should enable protection and restoration, making the network resilient to power failures. The basic topology consists of a single fiber collector ring with stars attached to it. The collector ring strings up RNs, which are the centers of the stars. The ONUs attached to the RN on the west side of the ring talk and listen to the transceiver on the west side of OLT, and likewise for the ONU attached to the RN on the east side of the ring. At a logical level, there is a point-to-point connection between each RN and OLT. When there is a fiber cut, all affected RNs will sense the signal loss and flip their orientation.

A RN has either a passive power splitter (coupler) or an AWG inside. If a RN contains a passive splitter, one dedicated wavelength on DWDM grid is used to broadcast the downstream data for the ONUs attached to the RN. Correspondingly, the ONUs have transmitters that send upstream data on CWDM grids. On the other hand, if a RN contains an AWG, each ONU has its own dedicated wavelength on a DWDM grid to communicate with the OLT. The downstream traffic and upstream traffic belonging to the same ONU may use the same wavelength, but different direction on the same fiber. Moreover, the authors propose a particular WDM PON MAC protocol but do not present any WDM DBA algorithm.

In the second project, the authors of [85] present the SUCCESS-DWA PON architecture that offers scalability by employing dynamic wavelength allocation (DWA) to provide bandwidth sharing across multiple physical PONs. The presented architecture requires that all wavelengths from the OLT reach all ONUs. Tunable lasers coupled together via a cyclic or nearly cyclic AWG and the AWG are implemented in the headend while WDM filters and a burst-mode receiver are employed within the ONUs. Each tunable laser is scheduled to communicate with several ONUs scattered over different ONU groups by sweeping its wavelength. The upstream traffic is separated from the downstream traffic with a wideband WDM filter between the AWG and the PON. On the other hand, incoming signals from the backbone network are separated into high-priority (HP) and best-effort (BE)traffic and then allocated to their own respective output queues to be assigned to appropriate tunable lasers. However, the presented model considers the implementation of only two traffic classes, which is not a sufficiently scalable solution for networks that transmit various multimedia applications.

The third project presents the SUCCESS-HPON as a next-generation hybrid WDM/TDM optical access architecture, based on a ring plus distribution trees topology, fast centralized tunable components, and novel scheduling algorithms [86]. According to the authors, the presented architecture includes the following:

- Backward compatibility—Guarantee for the coexistence of current-generation TDM PONs and next-generation WDM PONs in the same network;
- Easy upgradeability—Provide smooth migration paths from TDM PON to WDM PON;
- Protection/restoration capability—Support both residential and business users on the same access infrastructure.

The presented architecture of SUCESS-HPON includes a single-fiber collector ring with stars attached collector ring with stars attached to it. The collector ring strings up remote nodes (RNs), which are the centers of the stars. The ONUs attached to the RN on the one side of the ring talk and listens to the transceiver on that side of OLT, and likewise for the ONU attached to the RNs on the opposite side of the ring. There is a point-to-point WDM connection between the OLT and each RN. No wavelength is reused on the collector ring. When there is a fiber cut, affected RNs will sense the signal loss and flip their orientation.

The next architecture that involves the implementation of an AWG is given in [87]. The authors propose a hybrid WDM/TDM PON with wavelength-selection-free transmitters in which each upstream wavelength channel can be shared among multiple ONUs by means of TDM. Here, the ONUs can use their wavelength-selection-free (without wavelength tuning) transmitters to operate on any wavelength. The presented WDM/TDM PON.has a double-star topology serving the total of 128 subscribers in a cascade of 1:16 AWG and 1:8 power splitter at the remote node (RN). The authors define the 16 channels of the AWGs as wavelength channels and the 128 channels for subscribers as channels. The wavelength-selection-free transmitter is an un-cooled Fabry-Perot laser diode (FP-LD) wavelength-locked to an externally injected narrow-band amplified spontaneous emission (ASE). The wavelength-locked FP-LDs transmit 1.25 Gbps data to the subscribers over the feeder fiber where the feeder fiber is a standard single mode fiber.

Another WDM EPON. based on the AWG implementation is presented in [88]. The use of the AWG enables the proposed WDM EPON architecture to achieve wavelength spatial reuse and to provide intercommunication capability between ONUs. As a result, the architecture not only allows upstream access to the central office, but also facilitates a truly shared LAN capability among the end users. Furthermore, the proposed WDM EPON. scheme takes account of the requirement for backward compatibility with the IEEE 802.3ah MPCP and incorporates a DBA scheme and a QoS provisioning mechanism to arbitrate the access of the individual ONUs over the WDM layer. The proposed WDM EPON architecture has a

full-duplex tree topology, in which a dual-fiber is employed to facilitate simultaneous transmissions in the downstream and upstream directions, respectively. The
cyclic AWG not only enables the ONUs to access the upstream bandwidth, but also
allows them to communicate directly with one another. The AWG permits each
wavelength in the network to be spatially reused.

Upgrading the existing PON to the above mentioned WDM EPON systems
requires the replacement of the existing power splitter with an AWG router.
However, this upgrade is not particularly desirable as it requires work on the outside
plant and disrupts the existing customers. Moreover, having in mind that the
implementation of new architectures requires large investments, it is more likely
that future WDM EPONs will be primarily created as an upgrade to the conventional EPON.system, while the case of building a new WDM EPON infrastructure
in practice will be less frequent. This is the main reason why most authors are now
focusing on the bandwidth and resource management, access control, and QoS
algorithms rather than on the development of new architectures.

The authors of [74] formulated the problem of bandwidth management for
WDM EPONs and utilized scheduling theory (presented in the previous section).
The presented simulation results indicate that the scheduling used in a WDM EPON
can severely impact the network performances, and according to authors, the
optimal scheduler is a hybrid offline-online scheduler.

The authors of [89] propose a new WDM DBA scheme based on the Interleaved
polling with adaptive cycle time (IPACT), called WDM IPACT with a single
polling table (WDM IPACT-ST). Here, the IPACT technique is adopted and applied on multichannel WDM PON, where ONUs are equipped with an array of
fixed-tuned transceivers, one for each upstream/downstream wavelength channel. In
the proposed solution, transmission windows are assigned to ONUs in a
round-robin fashion, allowing them to transmit on the first available upstream
wavelength channel (first fit). To apply this algorithm, the OLT has to know which
upstream channel will first become idle. Consequently, the OLT has to monitor all
upstream channels and must calculate their next idle time, i.e., the OLT is able to
predict exactly when a particular upstream channel will become idle (OLT knows
the RTT of all ONUs and the size of the transmission window assigned to each
ONU for the transmission on upstream channels). Moreover, after it releases the
GATE messages to the ONUs, the OLT knows exactly when a transmission will be
finished on all upstream channels. Therefore, the OLT can schedule an ONU to
transmit on the upstream channel that has the smallest value of the idle time. The
length of the transmission window allocated to each individual ONU is assigned by
the OLT. The length of the window can be a predefined (fixed) value or it can be
based on some algorithms that use the request of each ONU as an input for the
calculation.

The authors analyzed the fixed, limited, and gated assignment schemes from
[47, 62]. The authors demonstrate that the resultant 2-channel and 3-channel WDM
IPACT-ST EPON. outperforms a single-channel TDM IPACT EPON in terms of
delay, regardless of the number of deployed ONUs (16, 32 or 64). Namely, in the
single-channel EPON., the polling cycle linearly increases with an increasing

number of ONUs, while in multichannel EPONs, the simultaneous use of multiple wavelengths accommodates an increasing number of ONUs while maintaining a short polling cycle. Furthermore, the authors analyze the QoS support through the implementation of strict priority. Nevertheless, both IPACT and strict priority scheduling proved not to have the ability to handle the excess bandwidth found from overloaded ONUs (i.e., ONUs requesting more than the minimum guaranteed bandwidth), and not to provide a fair and efficient intra-ONU bandwidth allocation.

A similar approach was introduced in [90]. The authors present two dynamic bandwidth allocation schemes that enable different ONUs to efficiently share the access network bandwidth within the time and wavelength domain—online and offline scheduling. In the online scheme, OLT makes a scheduling decision based on individual requests and without the global knowledge of the current bandwidth requirements of other ONUs (in a grant-on-the-fly manner). In offline scheduling, the ONUs are scheduled for transmission once the OLT has received current MPCP report messages from all ONUs, allowing OLT to take into consideration the current bandwidth requirements from all stations (a wait-for-all manner). Conversely, no excess bandwidth was taken into consideration and no QoS WDM DBA was presented.

In [70], authors propose a WDM IPACT scheme similar to [74], in which they use TDM and wavelength assignment concurrently to dynamically allocate bandwidth. They propose a method for extending IPACT to WDM, since IPACT, in its simplest form, essentially represents a round-robin statistical multiplexing scheduler. In the proposed scheme, instead of waiting for an ONU to complete its upstream transmission before moving to the next ONU, the system can instantaneously move to the next ONU that can be served by the next available supported wavelength. Contrary to [89], their solution supports different ONU architectures (each ONU will have its own subset of supported wavelengths and may consist of a tunable laser/receiver or a fixed array of laser/receivers). The QoS implementation was not presented in the above mentioned model.

In all of the previously discussed models, the channel selection is based on the first-fit technique (i.e., first available free channel).

The authors of [91] propose two different architectures along with different wavelength and bandwidth allocation schemes. The first architecture (scheme A1) assumes a fixed grouping of ONUs where the ONUs are divided into multiple subsets and each is allocated a fixed wavelength channel for upstream transmission. Accordingly, every ONU maintains a fixed transceiver, while the OLT maintains a bank of fixed transceivers. Within each subset, the transmission of different ONUs is arbitrated by the OLT through either a fixed or dynamic time-division slot assignment scheme.

The second architecture (scheme A2) is more flexible and allows for simultaneous time sharing and wavelength sharing. For the upstream transmission, every ONU can be equipped with one or more fixed transmitters, and every ONU informs, during the registration process, the OLT of the wavelength(s) it can support for appropriate resource allocation and management. The OLT, upon receiving bandwidth requests, allocates transmission windows for the various ONUs, taking into account the wavelengths they support. Alternatively, the ONU could optionally maintain a fast tunable laser to allow for more flexibility.

The first allocation scheme called the static wavelength dynamic time (SWDT) relies on the simple architecture of A1 and divides the set of all stations into subsets each of which share one wavelength. The OLT allocates wavelengths statically among ONUs, and the upstream bandwidth is assigned dynamically depending on the request of each ONU. Further, ONUs are divided into many classes which are proportionate to the number of wavelengths, and hence, each class will share a predetermined wavelength. Since the number of ONUs on each wavelength is identified, SWDT runs on each channel separately, i.e., the OLT waits until all reports from one subset are received and then runs the allocation algorithm. Although easy to implement, this scheme underutilizes network bandwidth since it does not exploit the inter-channel statistical multiplexing. Namely, when the load on one particular channel is light and high on another, the OLT cannot use the available bandwidth on that lightly loaded channel and reassign it to highly loaded ONUs on another wavelength, which would certainly result in better performance. This scheme does not allow dynamic channel allocation, but it does allow for DBA [73] on one particular wavelength.

The other presented schemes are based on the DBAs introduced in [55] and allocate excess bandwidth in two different ways in order to efficiently utilize the available WDM PON resources by enabling inter-channel statistical multiplexing. The authors discuss the problem of grant scheduling in the context of excess bandwidth distribution and present three different algorithms for dynamic resource allocation. All of the presented dynamic wavelength dynamic time (DWDT) schemes rely on the second architecture A2 and enable the dynamic allocation of bandwidth for different ONUs in both the wavelength and time domains. In the first algorithm DWDT-1, the OLT waits until all the REPORTs are received from all ONUs (on all channels). Upon that receipt, the OLT runs a bandwidth allocation algorithm to determine the bandwidth and channel for every ONU. The excessive bandwidth is calculated according to [55], and authors suggest and discuss three models for its allocation, namely the controlled (CE), fair (FE), and uncontrolled (UE) model of allocation.

In the UE scheme, the OLT collects all the excessive bandwidth available for the next cycle from the received REPORTs and assigns this total excess uniformly to all highly loaded ONUs, regardless of their requested bandwidth. Hence, if some ONUs are only 'slightly' highly loaded, they are being assigned an unfair share of the excess bandwidth that could ultimately not be utilized. Consequently, the assignment of the excess bandwidth must be controlled by the OLT in order to guarantee a fair bandwidth allocation for all highly loaded ONUs. In order to resolve this issue, the authors suggest the implementation of the CE scheme. The CE scheme allocates the excessive bandwidth in a round-robin fashion and hence some highly loaded ONUs might not have the chance to receive any share of the excessive bandwidth due to the fact that the total excessive bandwidth will be 'spent' before visiting all ONUs, or as the authors imply, these 'last' ONUs might get a less share than the 'first' ones. For that reason, the authors propose a fair-excess (FE) allocation scheme that assigns portions to highly loaded ONUs according to their bandwidth demand. As a result, FE will ensure fair-excess bandwidth allocation among all highly loaded ONUs.

Although the DWDT-1 enables ONUs to share network resources in both the time and wavelength domains and even though it improves system efficiency in comparison with the SWDT, this algorithm also has deficiencies which arise from the fact that algorithm is based on offline scheduling. In order to solve the stated efficiency problem, the authors further propose the implementation of the DWBA-2 and DWBA-3 algorithms. In these algorithms, the OLT sends GATE messages 'on the fly' to all ONUs requesting less than the minimum guaranteed bandwidth. This 'on the fly' bandwidth assignment mitigates the effects of the channel idle time experienced by the DWBA-1 and results in a better throughput and delay performance. However, such a scheme may increase the complexity of the design and implementation of the DWBA due to the fact that the OLT must keep track of each REPORT message received from each ONU to be able to assign the appropriate transmission window.

QoS support was discussed in [92], and the authors propose two approaches for providing QoS support in hybrid TDM/WDM EPON.. The first solution, namely the QoS-DBA-1, can be implemented in both A1 and A2 WDM PON architectures from [91]. In this scheme, in order to integrate QoS with DWBA-2 authors proposed the implementation of the M-DWRR intra-ONU scheduler. However, the presented simulation results show dependency on the adopted intra-ONU scheduler weight profile. In the second approach, the authors present two DWBA models, namely the QoS-DBA-2 and QoS-DBA-3 model. Both models rely on the WDM PON architecture A2 from [91] and in comparison with algorithms presented there, both the OLT and ONU are responsible for performing bandwidth allocation. Moreover, both models segregate the transmission of high-priority traffic from medium and low-priority traffic by wavelength. In the QoS-DBA-2, this segregation is strict, while the QoS-DBA-3 scheme permits each ONU to transmit lower priority traffic on the wavelength dedicated for the transmission of the high-priority traffic. However, none of the proposed models segregates the transmission of medium- and low-priority traffic class and both use one wavelength for the transmission of these traffic classes.

The authors of [93] propose a new byte size clock (BSC) reservation MAC scheme that not only arbitrates upstream transmission and prevents optical collisions, but also varies bandwidth according to demand and priority, reduces request delay using pre-allocation and delta compression, and handles the addition/reconfiguration of network nodes efficiently. The proposed protocol is backward compatible with APON and EPON, i.e., the new access scheme exploits both WDM and TDM to cater for both light and heavy bandwidth requirements and supports both Ethernet and ATM packets without segmenting or aggregating them. Additionally, the amount of pre-allocated bandwidth can be minimized using delta compression, which in turn reduces the latency due to the request and grant mechanisms. By using delta compression to compute the delta (difference) between packets and by transmitting only the delta instead of the original packets, the need for reserving a large number of data slots is avoided and the number of pre-allocated timeslots is kept to a minimum. However, in BSC, all nodes need to be synchronized, and the resultant TDM frame time structure does not comply with the IEEE 802.3ah [46].

The authors of [94] present the solution for enabling multicast transmission with color-free ONUs. They propose a WDM PON that provides conventional unicast data and downstream multicast function. At the OLT, for each WDM channel, a dual-drive Mach-Zehnder modulator (DDMZM) is used to generate a sub-carrier double-sideband differential-phase-shift-keying (DPSK) signal. All the central carriers are separated and subsequently modulated to deliver the multicast data, while the remaining sub-carrier DPSK signals carry the downstream unicast traffic. In the ONUs, part of the downstream unicast signal power is remodulated for upstream transmission, which enables source-free ONUs.

As previously explained, one of the main reasons why the WDM EPON currently receives a lot of attention as a solution for the first or last-mile problem is that it can provide high bandwidth with a low cost. Namely, with new applications and services, the QoS support in the access network has become the major concern with reference to supporting the triple-play infrastructure [95–98]. Despite the fact that the introduction of more complex models for dynamic bandwidth allocation and implementation of intra-ONU mechanisms is likely to improve the performance of different classes, single-channel EPON systems do not scale well with the increasing number of users and ONUs. Just as it has once been the case in EPON, today, with the dramatic increase in the number of end users and bandwidth requirements, the QoS issue once again emerges, this time in WDM EPON.

Until now, only a few authors have analyzed and discussed QoS support in WDM EPONs [91, 92]. However, the presented models do not segregate the transmission of medium and low-priority traffic class and use only one wavelength for their transmission. Moreover, the authors of [91], even though discuss traffic segregation, analyze the system behavior and performances with only two implemented wavelengths. Besides that they do not discuss the case of the highest ingress load, i.e., WDM EPON with 64 ONUs. On the other hand, the authors of [94] discuss the transmission of the multicast traffic as a major factor for the transmission of the multimedia applications in WDM EPON, but do not present any DWBA algorithm and do not discuss the QoS support and implementation.

In the following chapters, we analyze the WDM EPON architecture which reconsiders all of the above stated issues. Moreover, we present and analyze two system models with full QoS, namely the fixed wavelength priority bandwidth allocation (FWPBA) and the dynamic wavelength priority bandwidth allocation (DWPBA). Within the proposed models, we further analyze DWBA algorithms for WDM EPON and discuss their characteristics and implementation.

Chapter 6
WDM EPON Architecture

6.1 Introduction

Following the extensive discussion from the previous section, we may now conclude that WDMtechnologies can be used to dramatically increase the throughput of EPONs. However, the probability of constructing a new WDM EPON network is very small, considering the large investments that would follow the construction of such a system. We should bear in mind the cost of tunable and wavelength-sensitive WDM components, which is quite high for access networks.Moreover, in the access network, the cost of the implemented components is shared among a few tens of residential users in contrast to the long-haul networks, where implementation costs are shared among thousands of business users. Thus, the development of network architectures, subsystems, and devices that could reduce the cost of WDM EPONs represents a crucial factor for their successful deployment.

In light of these conclusions, different authors have lately focused on and extensively discussed the solutions for upgrading the single-channel system to a multichannel system as the cost of this upgrade would still be comparatively smaller than the cost required for the construction of a new WDM EPON system, i.e., laying new fiber either through digging new trenches or tunneling through the existing infrastructure. Moreover, this would allow service providers to make decisions about an upgrade in accordance with their actual needs and capacities in a 'pay as you grow' manner [71].

The implementation of the WDM technology in EPONs includes [71]:

- The implementation of an array of fixed laser/receivers in the OLT;
- The implementation of either an array of fixed laser/receivers or one or more tunable laser/receivers in the ONUs.

However, managing ONUs with different WDM capabilities would be extremely difficult because service providers would then have to record the types of ONUs deployed as well as maintain spare parts inventories for each of the components.

© Academic Mind and Springer International Publishing AG 2017   175
M. Radivojević and P. Matavulj, *The Emerging WDM EPON*,
DOI 10.1007/978-3-319-54224-9_6

Accordingly, it is more likely that service providers will use either tunable laser/ receivers or fixed laser/receiver arrays, not both.

In this chapter, we present and analyze two dynamic wavelength and bandwidth allocation (DWBA) models for the hybrid TDM/WDM EPON network which are in line with the upgrade strategy of EPONs to WDM EPONs. Both models, namely the fixed wavelength priority bandwidth allocation (FWPBA) and the dynamic wavelength priority bandwidth allocation (DWPBA), support QoS [95–98]. In the proposed models, we suggest and analyze DWBA algorithms for WDM EPON. Both models incorporate a novel approach in analyzing QoS in WDM EPON in which wavelength assignment takes place per service class and not per ONU, as it has been a common approach in literature so far. In this way, we avoid the need for the implementation of additional complex algorithms for QoS support, which in turn significantly reduces DBA complexity and directly decreases system cost in comparison with various QoS DBA models which have been published until now [95–98].

From the point of view of optical devices, it is expected that advanced WDM optical devices, capable of sending/receiving 1–10 Gbps per wavelength in addition to multiple wavelengths, will be simultaneously commercialized in the future [82]. Moreover, QoS support will become essential for the successful delivery of various real-time applications and services that demand bandwidth guarantees ranging from several Gbps to the estimated Tbps (online gaming, e-HealthE, 3D video, digital cinema, and many more). The presented 'wavelength per service class' approach simplifies the introduction of different traffic classes and allows service providers to offer end users a range of services without additional algorithms and the subsequent upgrade of equipment.

The proposed WDM EPON system is actually a tree-based EPON with support to multiple wavelengths $(\lambda_0, \lambda_1, \lambda_2, \lambda_3)$, in every ONU and OLT. Figure 6.1 shows a WDM EPON in which four different wavelengths have been supported in every

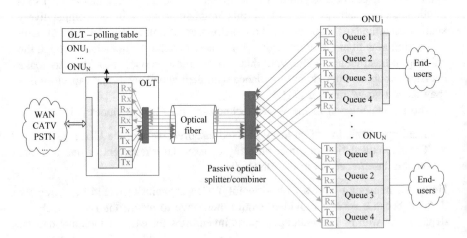

Fig. 6.1 WDM EPON architecture

unit. Hybrid TDM/WDM system allows simple migration from a single-channel TDM EPON to the multichannel WDM EPON system. The presented solution imposes a particular WDM architecture on the ONUs and enables the OLT to schedule transmission to and reception from ONUs on any wavelength supported by that ONU.

Headend is equipped with four fixed transmitters for simultaneous downstream transmissions and four fixed receivers that are constantly receiving data transmitted by stations in all upstream channels, Fig. 6.2. In the initial phase of migration, we analyze a solution with an array of fixed-tuned transceivers in the OLT and ONUs, one for each operating wavelength channel in order to enable simultaneous transmission of traffic in one station on different wavelengths. Furthermore, both models could be easily implemented in the existing EPON systems on a per need basis, i.e., only nodes with higher traffic demands may be WDM upgraded. With an array of fixed-tuned transceivers, every station is able to simultaneously use three different wavelengths and exchange REPORT and GATE MPCP control messages on the defined wavelength, Fig. 6.3. In the downstream direction, this would allow the OLT to send transmissions on multiple wavelength channels concurrently and would include extensions of MPCP for WDM EPON. With the implementation of tunable transceivers, optical units can only use a single wavelength channel at any given time, which means that the bandwidth currently available with a single fixed-tuned transceiver will not be expanded enough. Moreover, since the dead tuning time must be imposed every time, there is a wavelength switch, this actually provides less bandwidth [71].

In the presented models, one of the supported wavelengths (λ_0) is reserved for discovery services and synchronization because the OLT becomes aware of ONU's WDM architecture after the discovery and the registration process. This wavelength can be either the original wavelength that has been defined for EPON usage or some

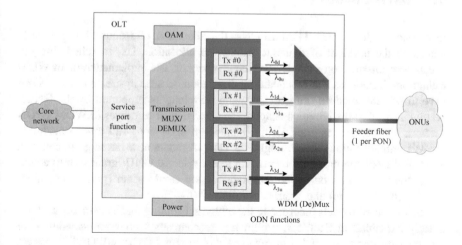

Fig. 6.2 OLT logical architecture

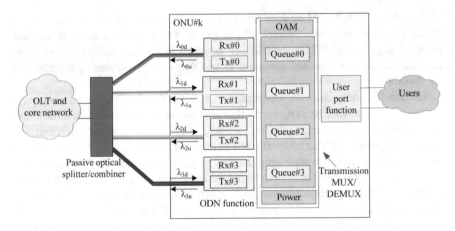

Fig. 6.3 ONU logical architecture

other predefined wavelength. For backward compatibility, we suggest the implementation of the original wavelength channel of TDM EPONs. Other three wavelengths are used for data transmission according to the dynamic wavelength and bandwidth allocation algorithms presented below. We suggest that four supported wavelengths should be selected from the C band, since a wide range of equipment supports this band and it is more appealing to vendors (Sect. 5.2).

In the following sections, we present, analyze, and compare the suggested models and algorithms in terms of delay, jitter, queue occupancy, throughput, and overall system performance.

6.2   DWBA Schemes

In classic single-channel TDM EPONs, the allocation of resources is primarily related to the problem of dynamic bandwidth allocation, i.e., to scheduling upstream transmissions on a single wavelength channel. The implementation of WDM technology introduces a new scheduling dimension because upstream transmissions have to be scheduled on different upstream wavelengths supported by the ONUs. As emphasized in the previous chapter, the allocation of resources in WDM EPON network can be seen as grant sizing (bandwidth allocation) and grant scheduling (wavelength allocation). Accordingly, the DWBA model now has to grant a certain amount of available upstream bandwidth to the selected ONU (grant sizing) and to define the wavelength on which that transmission would occur (grant scheduling) (bandwidth allocation) and grant scheduling [74].

The presented models incorporate the offline scheduling mechanism. As we have already explained in Sect. 5.7, within this mechanism, ONUs are scheduled for transmission once the OLT has received the current MPCP REPORT messages from all ONUs. This mechanism allows the OLT to take into consideration the current bandwidth requirements of all ONUs and to use them for the scheduling and grant sizing, i.e., the scheduling pool contains all ONUs. Furthermore, we present a new approach to QoS implementation in WDM EPON in which we use wavelength assignment per service class to improve service sensitivity and customer subjective impression [95].

The QoS support is implemented in accordance with the DiffServ model [60] through the introduction and support of the following traffic classes:

- EF (Expedited Forwarding)—highest priority traffic class for delay-sensitive traffic with constant bit rate which is intended for services such as voice and other delay-sensitive applications that require bounded end-to-end delay and jitter specifications;
- AF (Assured Forwarding)—medium-priority traffic for not delay-sensitive traffic with variable bit rate. AF class is intended for services, such as video transmission, that are not delay-sensitive but which require bandwidth guarantees. Assured forwarding allows the operator to provide assurance of delivery as long as the traffic does not exceed a certain subscribed rate. Traffic that exceeds the subscription rate faces a higher probability of being dropped if congestion occurs;
- BE (Best-Effort)—low-priority traffic class for delay-tolerable services that do not require any guarantees in terms of jitter and bandwidth, such as web browsing, file transfer, and e-mail applications.

In order to support differentiated services over WDM EPON, each ONU is installed with an array of physical queues, where each queue is used to store a defined class of traffic, Fig. 6.3. When a packet arrives to ONU, it will be first categorized into one of the priority groups according to its content and then it will be placed into one of the queues. Furthermore, in both models, we segregate the transmission of different traffic classes by wavelengths in the following way:

- λ_1 wavelength is allocated for transmission of highest priority EF traffic class;
- λ_2 wavelength is allocated for transmission of medium-priority AF traffic class;
- λ_3 wavelength is allocated for transmission of lowest priority BE traffic class.

Having segregated the transmission of traffic classes, there is no need for the implementation of an additional algorithm for QoS support. The presented models are completely in compliance with the IEEE 802.3ah standard and incorporate the previously mentioned offline scheduling in order to support the QoS implementation. With the aim of supporting different wavelengths and future developments of the presented models, we have implemented the extensions of MPCP for WDM EPON [71]. The MPCP GATE message is modified by means of an additional field (one byte) indicating the channel number assigned by the OLT to the ONU. Since both models presented below support the previously explained extension of the MPCP protocol, they could be upgraded in terms of the number of supported wavelengths and stations on a per need basis.

6.3   FWPBA Model

FWPBA model enables the transmission on four different wavelengths and involves the previously described architectures of the OLT and ONUs. One wavelength (λ_0) is reserved for the synchronization and transmission of control protocol messages, while the three other wavelengths transmit the reserved traffic classes. Each data wavelength is strictly associated with a single traffic class and can only be used for the transmission of such traffic class [95–97].

In case of the implementation of the FWPBA model, communication within the system implies that the FWPBA model should make a fixed link between the type of traffic and the wavelength which is known to the OLT and is used for the transmission of such traffic. In this way, wavelength allocation is not required, and the overall result is that the algorithm appears to be more efficient.

In the downstream direction, by means of the broadcasting mechanism, the OLT sends data to the ONUs simultaneously on multiple wavelengths, which are in turn received by the ONUs based on the destination MAC address (as is the case with a classic EPON [42]). In the upstream direction, and in accordance with MPCP protocol and rules of the offline scheduling mechanism, every ONU generates a REPORT message, one for each of the supported traffic classes, where each message contains the bandwidth requirement for the corresponding class, i.e., queue occupancy. Following the arrival of REPORT messages from all ONUs, the DBA module for the allocation of bandwidth in the OLT performs bandwidth allocation for each traffic class in all ONUs in accordance with the mathematical model presented below. In each cycle, every ONU transmits traffic concurrently on three different wavelengths and retains all three wavelengths until the process of transmission on each of them has been finalized.

Given the fact that one ONU retains all wavelengths until the transfer on the most loaded wavelength has been completed, the module for bandwidth allocation allocates the same amount of available upstream bandwidth to each ONU based on the maximum requested bandwidth. Now, the OLT generates a single GRANT MPCP message with the approved amount of bandwidth for all three traffic classes, Fig. 6.4.

One wavelength transmits only one defined traffic class. When the upstream transmission of all traffic classes is finished, ONU releases wavelengths and the next scheduled station can transmit the traffic (Fig. 6.5). Therefore, the most loaded wavelength dictates the duration of the transmission within the observed unit. Consequently, wavelengths that occupy minor bandwidth have to wait until the transmission on the most loaded wavelength is over. This introduces new waiting component, further referred to as t^{waiting}.

In order to analyze the characteristics of the model, we will assume that AF traffic occupies the major part of the bandwidth in the system, i.e., λ_2 is the most loaded wavelength. Since the most loaded traffic class or the traffic class with the largest bandwidth demand defines the duration of the transmission in ONU, AF class will not have the additional waiting time as shown in Fig. 6.5. ONUi retains

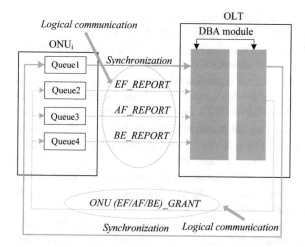

Fig. 6.4 Communication between OLT and ONU$_i$ in the FWPBA model

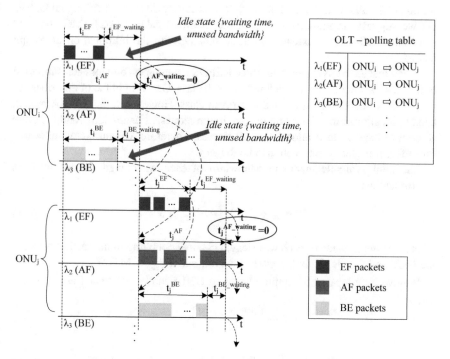

Fig. 6.5 Upstream ONU transmission in FWPBA model

all wavelengths until transmission on most loaded λ_2 has been executed. In the meantime, λ_1 and λ_3 are in the idle state. When the transmission of AF traffic class is over, ONU$_i$ releases wavelengths and the OLT schedules them for transmission in ONUj in accordance with the polling table.

It is obvious that the most loaded wavelength or traffic class with the largest bandwidth demand, in this case AF, will not have the additional waiting time. Moreover, the fact that the DBA module allocates the same bandwidth for all three wavelengths based on the maximum requested bandwidth introduces an additional inefficiency within the system—unused bandwidth in case of the two less loaded wavelengths. Namely, for both EF and BE classes, the OLT allocates more bandwidth than they have actually requested, and that bandwidth is not fully utilized. On the other hand, the simplification and reduction of the processing time of the OLT make the system more efficient.

The DBA algorithm used in both models is based on the gated IPACT scheme [42, 47]. This scheme will grant ONU a transmission window whatever size it has previously requested but not larger than the defined queue size. Thus, the largest possible granted window size will be the maximum length of ONU's queue. Hence, we do not assume that the total network load is evenly distributed among all ONUs, yet we suggest the implementation of the modified gated IPACT (MG- IPACT) algorithm. MG-IPACT extends the basic scheme with the introduction of weight factors.

Given that each ONU generates three REPORT messages, one for each supported traffic class, a weight factor is defined for each class. One weight factor describes the proportion of bandwidth allocated to a given class. However, in the FWBA model, bandwidth allocation is based on the maximum requested bandwidth , and the same amount of bandwidth is allocated to all classes. Therefore, weight factors allocated for different traffic classes will also be the same.

The total available upstream bandwidth of each wavelength in one cycle is determined as:

$$W^{\text{total}} = R * \left(T^{\text{max\_cycle}} - N * T_g\right), \tag{6.1}$$

where R is the line rate of each wavelength channel operating in the section between the OLT and ONUs, T_g is the guard interval, and $T^{\text{max}}_{\text{cycle}}$ is MTCT.

Bandwidth allocated for traffic classes in ONUi can be calculated as:

$$W_i^{\text{max\_requested}} = \max\left\{W_i^{\text{tc\_requested}}\right\}, \quad \text{tc} \in \{\text{EF}, \text{AF}, \text{BE}\}, \tag{6.2}$$

$$W_{\text{total}}^{\text{tc\_requested}} = \sum_{i=1}^{N} W_i^{\text{max\_requested}}, \quad \text{tc} \in \{\text{EF}, \text{AF}, \text{BE}\}, \tag{6.3}$$

$$w_i^{\text{tc}} = \frac{W_i^{\text{max\_requested}}}{W_{\text{total}}^{\text{tc\_requested}}}, \quad \text{tc} \in \{\text{EF}, \text{AF}, \text{BE}\} \quad \text{and} \quad \sum_{i=1}^{N} w_i^{\text{tc}} = 1, \tag{6.4}$$

$$W_i^{tc\_allocated} = \begin{cases} w_i^{tc} * W^{total}, & W_i^{tc\_requested} < W^{queue} \\ W^{queue}, & W_i^{tc\_requested} \geq W^{queue} \end{cases}, \quad tc \in \{EF, AF, BE\}, \quad (6.5)$$

where:

W^{total}	The total available upstream bandwidth of each wavelength in one cycle $W_i^{tc\_requested}$ bandwidth requested for each traffic class in ONU$_i$,
w_i^{tc}	the weight assigned to ONU$_i$ for each supported traffic class, where $\sum_{i=1}^{N} w_i^{tc} = 1$
N	number of ONUs in the system,
$W_i^{tc\_requested}$	bandwidth requested for each traffic class in ONU$_i$,
$W_i^{max\_requested}$	maximum requested bandwidth of all three traffic classes in ONU$_i$,
$W_{total}^{tc\_requested}$	total requested bandwidth of all ONUs for each traffic class,
$W_i^{tc\_allocated}$	bandwidth allocated for each traffic class in ONU$_i$,
W^{queue}	maximum defined length of ONU's queue

As we have previously emphasized, the FWPBA model is simple and efficient but suffers from two shortcomings—additional delays and unused bandwidth. The rest of this section presents the analysis of the influence of the aforementioned components on system performance.

Since the wavelength allocation is static in the FWPBA model, this means that one station within the system transmits traffic on three wavelengths concurrently and retains all three wavelengths until transmission on all of them has been executed. This behavior introduces an additional waiting time ($t^{waiting}$), which can be expressed as (Fig. 6.5):

$$t_i^{ONU} = \max\{t_i^{tc}\}, \quad tc \in \{EF, AF, BE\}, \quad (6.6)$$

$$t_i^{tc\_waiting} = t_i^{ONU} - t_i^{tc}, \quad tc \in \{EF, AF, BE\}, \quad (6.7)$$

where t_i^{ONU} represents the processing time in ONU$_i$, $t_i^{tc\_waiting}$ and t_i^{tc} represents the waiting and transmission time of traffic classes in ONUi, respectively.

According to (6.6) and (6.7) and based on the previous discussion, the most loaded wavelength does not have this additional waiting time. Since AF traffic class is the most present traffic class in the system, this class does not have the waiting component either. Given the fact that multimedia traffic is nowadays dominant, we assume that this is the most likely situation in practice. At the same time, EF traffic class usually occupies the minimum amount of available bandwidth due to the fact that voice packets that belong to this class are small, which indicates that the waiting time will be dominant for this traffic class.

Because of the reasons stated above, the FWPBA algorithm for bandwidth allocation introduces an additional inefficiency in system. The DBA module in the OLT allocates the same value for all three traffic classes and, as presented in

(6.2)–(6.6), the decision is made based on the maximum requested bandwidth for all three classes. Consequently, OLT performance values are better, but the allocated bandwidth is totally utilized only on a most loaded wavelength. The amount of unused bandwidth for each traffic class in ONUi can be expressed in the following way:

$$W_i^{tc\_unused} = W_i^{tc\_allocated} - W_i^{tc\_requested}, \quad tc \in \{EF, AF, BE\}, \quad (6.8)$$

$$W_i^{ONU\_unused} = \sum_{tc \in \{EF, AF, BE\}} W_i^{tc\_unused}, \quad (6.9)$$

where $W_i^{tc\_unused}$ represents the unused portions of available bandwidth for each traffic class in ONU$_i$, and $W_i^{ONU\_unused}$ represents the total unused bandwidth allocated for ONU$_i$.

Based on our assumption related to the traffic profile, the AF traffic class, as the most present traffic class in the system, will not have the unused bandwidth component. At the same time, other two data wavelengths (in this case λ_1 and λ_3 that transmit EF and BE traffic classes) would be in idle state and underutilized, since the OLT cannot use the available bandwidth on the lightly loaded channel and reassign it to a highly loaded channel. Although simple and easy to implement, the FWPBA scheme underutilizes network resources and, from now on, we will use this scheme as a basis for our comparative study.

6.4   DWPBA Model

The principle applied in the FWPBA model, in which an ONU retains all three wavelengths until the transmission on all three wavelengths is finished, does not use system resources efficiently.

In order to improve system performance, we abandon the described mechanisms [95, 98]. In the presented DWPBA model, the wavelength on which an ONU has finished transmission of the defined traffic class is immediately released and allocated to the next ONU for the transmission of the same class in accordance with the OLT polling table. One wavelength could be used only for the transmission of the defined traffic class, just as it was the case in the FWPBA model.

The data exchange within the model is similar to that of the FWPBA model. In the DWPBA model, we have defined the same static mapping between traffic classes and data wavelengths in order to avoid the implementation of an additional algorithm for wavelength allocation. However, the exchange of control messages is the same as the exchange defined for the first model which was presented. In the downstream direction, the OLT, using a broadcast mechanism, sends data to ONUs changing data wavelengths simultaneously, just as it has been the case with a single-channel EPON. In the upstream direction, all ONUs use REPORT MPCP message to request bandwidth for each traffic class independently. Having received REPORT messages from all ONUs, the DBA module in the OLT allocates bandwidth to each ONU and to each traffic class and notifies them by using a GRANT MPCP message (offline mechanism), Fig. 6.6.

Contrary to the FWPBA model, at the beginning of the cycle, the DBA model in the OLT allocates three data wavelengths to different ONUs based on the received bandwidth requests and according to the mathematical model presented, Fig. 6.7. In this model, wavelengths are further processed independently and in accordance with the polling table defined in the OLT. Consequently, dynamic bandwidth allocation is more complex in the DWPBA model in comparison with the FWPBA model because every wavelength is processed independently. Moreover, once again in contrast with the FWPBA model, the OLT now allocates a certain amount of available upstream bandwidth to each traffic class and accordingly generates three grant messages instead of one, as it has been the case in the previously described model, Fig. 6.6

When the transmission on one wavelength in the selected ONU is over, that wavelength is scheduled to the next ONU for the transmission of the same (defined) traffic class. The allocation of wavelengths is completely asynchronous, and at one point assigned to different wavelengths of ONU units: $ONUi$, ONU_g, and ONU_f for the transmission of EF, AF and BE classes, respectively, Fig. 6.7. When the transfer is complete, ONUs immediately release the appropriate wavelength for the given traffic class and the OLT assigns it to the next ONU. In the situation illustrated in Fig. 6.7, when the transfer of the EF traffic class is complete on the wavelength reserved for this class (λ_1), this wavelength is rescheduled to $ONUj$ for the transfer of the same class. At the same time, the transfer of the AF and BE traffic classes is independent and placed at defined wavelengths $(\lambda_2$ and $\lambda_3)$ in the ONU_g and ONU_f,

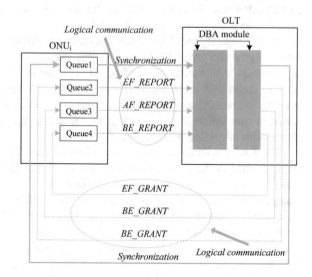

Fig. 6.6 OLT—ONU communication in DWPBA

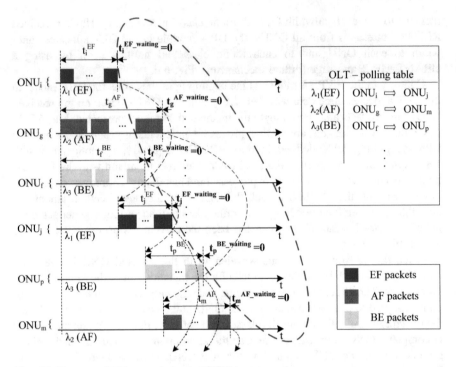

Fig. 6.7 Upstream ONU transmission in DWPBA

respectively. The λ_2 wavelength is next scheduled for the transmission of AF traffic class in ONU$_m$ according to the polling table in the OLT, λ_3 is rescheduled to ONU$_p$, and so on. Obviously, the waiting times do not exist any longer because the least loaded wavelengths in ONU are immediately shifted for transmission to other units in accordance with the polling table.

In this model, the OLT must maintain the information about the allocated wavelengths for each ONU in one cycle because every ONU must receive the opportunity to transmit each traffic class. The use of the independent processing of traffic classes and wavelengths further increases the complexity of the modules in the OLT. Therefore, different traffic classes are no longer synchronized and allocated to one station simultaneously, which means that we must use the modified bandwidth allocation scheme, Fig. 6.7.

The total available upstream bandwidth of each wavelength in one cycle has been defined in the same way as for the FWPBA model. As explained earlier, the calculation of the allocated bandwidth for traffic classes and weight factors is no longer based on the maximum requested bandwidth for all three traffic classes. Accordingly, the allocated bandwidth for traffic classes in ONUi can be calculated in the following way:

$$W_{\text{total}}^{\text{tc\_requested}} = \sum_{i=1}^{N} W_i^{\text{tc\_requested}}, \quad \text{tc} \in \{\text{EF}, \text{AF}, \text{BE}\}, \tag{6.10}$$

$$w_i^{\text{tc}} = \frac{W_i^{\text{tc\_requested}}}{W_{\text{total}}^{\text{tc\_requested}}}, \quad \text{tc} \in \{\text{EF}, \text{AF}, \text{BE}\} \quad \text{and} \quad \sum_{i=1}^{N} w_i^{\text{tc}} = 1, \tag{6.11}$$

$$W_i^{\text{tc\_allocated}} = \begin{cases} w_i^{\text{tc}} * W^{\text{total}}, W_i^{\text{tc\_requested}} < W^{\text{queue}} \\ W^{\text{queue}}, W_i^{\text{tc\_requested}} \geq W^{\text{queue}} \end{cases}, \quad \text{tc} \in \{\text{EF}, \text{AF}, \text{BE}\}, \tag{6.12}$$

where:

W^{total}	bandwidth requested for each traffic class in ONU_i,
N	number of ONUs in the system,
$W_i^{\text{tc\_requested}}$	bandwidth requested for each traffic class in ONU_i,
$W_{\text{total}}^{\text{tc\_requested}}$	total requested bandwidth of all ONUs for each traffic class,
$W_i^{\text{tc\_requested}}$	bandwidth allocated for each traffic class in ONU_i,
W^{queue}	maximum defined length of ONU's queue.

In the DWPBA model, the MG-IPACT algorithm allocates bandwidth for each of the traffic classes separately, and therefore the problem of unused bandwidth, as well the waiting component, is no longer an issue. However, even though a more complex DBA algorithm slightly degrades system performance in the control plane, we expect a much better use of available upstream bandwidth for data transfer in each wavelength channel.

In the next chapter, we conduct the detailed simulation experiments in order to verify our conclusions and validate the effectiveness of the proposed models.

6.5 Future Model Development

Until today, different models for incremental migration from TDM to TDM/WDM EPON.networks have been proposed, but most attention has been given to those solutions that support QoS implementation. Hence, with the rapid development of different bandwidth applications, QoS support is becoming a key concern in the WDM EPON network, just as it has similarly been the case with EPON networks. The future optical network must be able to support and guarantee low latency for high-speed, wide bandwidth applications that require several Gbps to Tbps speeds. Consequently, the bandwidth demands of end users will expand to 10 Gbps [99]. Under this assumption, a new concept for optical access networks will be needed.

The enhanced dynamic wavelength priority bandwidth allocation with fine scheduling (DWPBA-FS) model includes a hybrid inter/intra-ONU scheduling mechanism, where both the OLT and ONUs are responsible for performing packet scheduling [100,101]. The model includes an extension of the MPCP, thus allowing the presented architecture and models the possibility of an incremental upgrade from TDM EPON to TDM/WDM EPON based on the need. Moreover, with the multiple wavelength assignment for each ONU, it offers great flexibility and allows service providers to respond to the increase in user demands in terms of bandwidth and other QoS parameters in a very simple and cost-effective manner. This 'on demand' concept and the capability of the system to handle diverse user demands are currently considered to be a key factor for the realization of the new multi-service access network [102].

The analyzed WDM EPON system is a tree-based EPON with support to multiple wavelengths $(\lambda_1, \lambda_2, \lambda_3, \lambda_4)$ in every ONU and OLT, just as it has been the case in the previous DWPBA analysis. The OLT and ONUs architecture remains the same and is illustrated in Figs. 6.1, 6.2 and 6.3. As we can see in these Figures, there is an array of fixed-tuned transceivers in the OLT and ONUs, one for each operating wavelength channel, and their function is to enable simultaneous transmission of traffic in one station on different wavelengths. In other words, with the aid of fixed-tuned transceivers, every station is able to simultaneously use four different wavelengths and exchange REPORT and GATE MPCP control messages.

The DWPBA-FS model further develops the DWPBA model with the introduction of new traffic subclasses in the system in accordance with the Diffserv model [60]. As opposed to the FWPBA and DWPBA models, where the control wavelength is used exclusively for the synchronization and exchange of control messages, in the DWPBA-FS model, this control wavelength is now used for data transmission. The exchange of control messages in the enhanced model now takes place on data wavelengths, where the processing is carried out in accordance with the MPCP protocol. The QoS support is again realized through the support of differentiated services, i.e., the system's ability to support the transmission of different traffic classes. In this model, traffic classes are slightly different in comparison with the previous models. Namely, in the DWPBA-FS, network traffic is categorized in the following way:

- EF (Expedited Forwarding)—highest priority traffic class for delay-sensitive traffic with constant bit rate (Sect. 6.2);
- AF (Assured Forwarding)—medium-priority traffic for not delay-sensitive traffic with variable bit rate. The AF traffic class is further divided into four subclasses according to the standard (traffic is listed by priority, from highest to the lowest):

  - AF4 (e-commerce applications);
  - AF3 (mission critical application);
  - AF2 (non-organization streaming audio and video), and
  - AF1 (bulk traffic);

  The first two AF subclasses could be defined as premium and normal business applications, and another two as premium and home business applications. If the congestion occurs between classes, the traffic in the higher subclass is given priority over the lower traffic subclasses;

- BE (Best-Effort)—low-priority traffic class for delay-tolerable services that do not require any guarantees in terms of jitter and bandwidth (Sect. 6.2).

With the aim of reserving bandwidth for the four AF subclasses, we propose the implementation of the WFQ (Weighted Fair Queuing) as a scheduling algorithm in ONUs, since this method is able to automatically smooth out the flow of data by sorting packets which ultimately minimizes the average latency and improves system performances. Also, the WFQ can be described as a queuing algorithm that combines fair queuing and preferential weighting. The fairness, aspect of the WFQ, behaves in a way which is similar to the round-robin queuing, with queues serviced in a continuously repeating sequence from top to bottom, and then starting at the top again. The weighting aspect of the WFQ applies a 'weight' to a queue that indicates the importance of the queue in relation to the available resources. The weight is used to ensure that more important queues are processed more often than other less important queues. With the WFQ, queues are first sorted in the order of their increasing weighted value. Then, each queue is serviced according to its weighted proportion to the available resources. By using the WFQ, each priority group can reserve different weighted proportions in the next transmission window and the packet scheduling mechanisms allow different sessions to have different service shares [74]. Therefore, when the next GATE message arrives, the ONU can transmit packets from priority queues up to the amount that has previously been reserved.

Accordingly, we adopted the following scheme for the transmission and segregation of different traffic classes by wavelengths: λ_1 is allocated for the transmission of EF traffic class, λ_2 and λ_3 are allocated for the transmission of the AF traffic class, i.e. AF1, AF2, AF3 and AF4 traffic subclasses, and λ_4 is allocated for the transmission of the BE traffic class. For the transmission of medium-priority subclasses, we suggest the use of two wavelengths in one ONU simultaneously, since we assume that AF traffic class is the most present in the system. Namely, according to the new traffic scheme in modern networks, most applications and services are multimedia-based and in years to come, the different forms of video applications will account for approximately 90% of consumer traffic [103]. On the other hand, the EF traffic class, although being the highest priority class, includes voice applications that are not bandwidth intensive (10–20% of total generated traffic in the system) and with the allocation of one wavelength solely for its transmission, even the most rigorous QoS requirements can be met. BE traffic class generally occupies more bandwidth in comparison to EF class but, as we have previously explained, applications in this class do not require any QoS guarantees.

The DWPBA-FS model ensures fairness among all traffic classes, especially for BE traffic which, in most of the 'to-date' public schemes, pays the performance penalty to gratify higher priority traffic. With the allocation of one wavelength for this class, transmission characteristic of this class will be significantly improved in relation to different models which have been suggested so far. Having segregated the transmission of traffic classes in the described manner, there is no need for the implementation of an additional algorithm for QoS support.

The presented model is completely in compliance with the IEEE 802.3ah standard and introduces offline scheduling in order to support the QoS implementation. Scheduling in ONUs is defined in the following manner. Each ONU will follow a request-per-class trend to report its bandwidth requirements of each traffic class separately. Hence, each ONU will send three REPORT messages: one for the EF traffic (λ_1), one for all AF subclasses (λ_2), and one for the BE class (λ_4). Figure 6.8 illustrates the proposed intra-ONU scheduler. Since ONUs have the ability to transmit packets on λ_2 and λ_3 simultaneously, the WFQ-based scheduler will first schedule packets from Q4 to Q3 queues in accordance with the allocated bandwidth and the defined weight factors. At the same time, packets from lower priority queues (Q2 and Q1) will be scheduled for transmission on λ_3. In case the transmission window (TW) is still available and Q4 and Q3 queues are empty, or in case the remaining packets do not fit the TW (fragmentation is not allowed), the scheduler will try to schedule packets that belong to the AF2 and/or AF1 traffic classes in the remaining TW and vice versa. In this way, the scheduling mechanism efficiently uses system resources and adjusts the current distribution of traffic in the system in accordance with user demands.

Communication and data exchange between the OLT and ONUs are defined as follows. In the downstream direction, by means of the broadcasting mechanism, the OLT sends data to the ONUs simultaneously on multiple wavelengths, which are in turn received by the ONUs based on the destination MAC address (as is the case with the conventional EPON). However, in the upstream direction, apart from the bandwidth allocation, the allocation of resources now also includes wavelength allocation. At the beginning of the cycle, the OLT allocates four data wavelengths to the selected ONUs based on the received bandwidth requests and the mathematical model described below. When one ONU finishes transmission on the allocated wavelength(s), that wavelength(s) is (are) automatically shifted to another ONU for the transmission of traffic that belongs to the same class according to

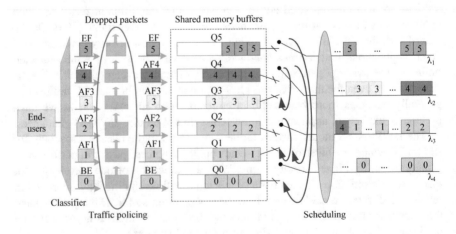

Fig. 6.8 Intra-ONU scheduling mechanism

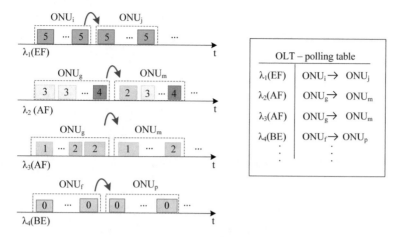

Fig. 6.9 The upstream ONU transmission in the DWPBA-FS model

decision made in OLT, Fig. 6.9. Accordingly, due to the fact that in one cycle every ONU must receive an opportunity to transmit the traffic of each traffic class, the OLT must maintain information about allocated wavelengths and ONUs which used them. Since wavelength allocation is fixed, the OLT has to make a decision only about bandwidth allocation, which significantly simplifies DBA complexity and improves system performances.

The DBA algorithm used in the DWPBA-FS model is based on the MG-IPACT algorithm used in its predecessor, i.e. the DWPBA model (6.10)–(6.12). As we have previously presented, the algorithm allocates the portion of the available bandwidth to supported traffic classes, i.e., to the EF, AF and BE traffic classes. A portion of the available bandwidth assigned by MG-IPACT to the AF traffic class is allocated to AF subclasses based on the defined weight factors of the WFQ mechanisms:

$$W_i^{AF_j\_allocated} = w_{q\_j} \times W_i^{AF}, \quad j \in \{1, 2, 3, 4\}, \tag{6.13}$$

where $W_i^{AF_j\_allocated}$ represents the bandwidth allocated to each AF traffic subclass, and w_{q_j} presents the weights (weights of the each queue which is defined for the transmission of AF subclasses), where $\sum_{j=1}^{4} w_{q\_j} = 1$.

Chapter 7
Performance Evaluation

7.1 Model Simulation

In this chapter, we study the performance of the presented FWPBA and DWPBA models and analyze the impact of the implementation of different traffic profiles on network performance. In addition, this section presents the results of the comparison of all key parameters of the respective models as well as the results of the comparison between the presented models, and significant DWBA models that have until now been presented in the literature within the field of QoS implementation in WDM EPON. In order to confirm the performance of a multichannel system, we further present the analysis and the comparison of the given model with the single-EPON system that supports QoS. Moreover, we present and evaluate the performances of the DWPBA-FS model as the next step in the development of the NGN optical access networks.

The presented FWPBA and DWPBA models are tested on a WDM EPONnetwork model developed in MATLAB where a Simulink packet has been used [104]. System modeling with MATLAB and the Simulink packet allowed us to implement and test the presented algorithms within the real network environment, just as it would be the case in the triple-play service provider's network.

In our simulations, we test a WDM EPON which consists of 16 or 64 WDM ONUs that support traffic transmission on four different wavelengths per each ONU. The speed of each wavelength channel amounts to 1 Gbps. The traffic load of stations varies between 0.1 and 1 (i.e., 10 and 100 Mbps), and the buffering queue size is 1 Mbytes. The maximum transmission cycle time is 2 ms and guard interval is set to 1 μs. For our simulations, the RTT was randomly generated in accordance with uniform distribution U[50, 200 μs], which corresponds to distances from ONUs to the OLT and amounts to approximately 15–30 km. When requesting the next time slot, a station must take into account the additional overhead that includes a frame preamble (8 bytes) and an interframe gap (12 bytes) between consecutive frames. Beside the RTT in our simulation, we further take into account the queuing

© Academic Mind and Springer International Publishing AG 2017
M. Radivojević and P. Matavulj, *The Emerging WDM EPON*,
DOI 10.1007/978-3-319-54224-9_7

delay, transmission delay and packet processing delay. The communication quantities used for the comparison are the following: average packet delay, maximum packet delay, queue occupancy, jitter, and packet loss rate.

An extensive study shows that most network traffic can be characterized by self-similarity and long-range dependence [64]. This model is used to generate highly bursty BE and AF traffic classes, and packet sizes are uniformly distributed between 64 and 1518 bytes. High-priority traffic is modeled by Poisson distribution with a packet size fixed to 70 bytes [48].

As referred to in the introduction section, the development of different applications and multimedia services is accompanied by new demands regarding network resources, and above all, demands in terms of the available bandwidth. With the aim of providing a detailed analysis of the performance of the presented model, we introduce the following traffic profiles—P1 and P2. The implemented profiles are as follows:

- Profile 1 (P1): EF (15%), AF (50%) and BE (35%).
- Profile 2 (P2): EF (20%), AF (40%) and BE (40%).

With the development of different bandwidth-intensive applications and their increasing popularity, the traffic schemes in next generation networks have been changed [1]. Since most popular multimedia and business applications could be categorized as medium priority traffic, i.e., as AF traffic class, this class is now dominant. Consistently, in the first traffic profile (P1): 15% of the total generated traffic is considered to be high priority, 50% is considered to be medium priority multimedia trafficof AF class, and the remaining 35% is considered to be low-priority BE traffic class [95–98, 100–101]. The second profile has been used in the majority of works that analyze the implementation of QoS in EPON and WDM EPON networks. With the aim of obtaining comparative results, we compare the second profile with the first profile (which we believe represents the future) and further provide the comparison of these models with different models which have been published until now.

7.2 FWPBA Model Analysis

A. *Average packet delay*

First of all, we analyze the average packet delays and waiting time components in the WDM EPONsystem with the FWPBA mechanism. Figures 7.1 and 7.2 illustrate the comparison between average packet delays and waiting time components for each of the supported traffic classes in the event of the implementation of both profiles. The simulation results confirm the conclusions obtained by the theoretical analysis of the model given in Sect. 6.3. Given that EF traffic class is the least presented class in the system, the waiting time for this class is dominant in

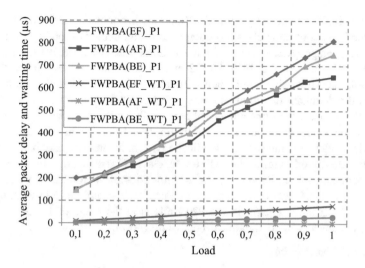

Fig. 7.1 Average packet delay and waiting time in FWPBA_P1 model

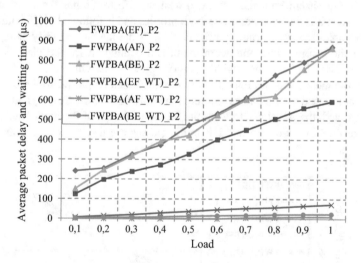

Fig. 7.2 Average packet delay and waiting time in FWPBA_P2 model

comparison with the BE and AF traffic, and high-priority class will have the highest average delay in the system in case of the implementation of both profiles. BE traffic class is more prevalent than EF traffic class and has a slightly better delay and waiting time characteristics. As expected, the average packet delayof EF and BE traffic classes at the highest load in FWPBA_P1 model is by 6.9 and 13% less than the delay in the FWPBA_P2 model because the amount of these traffic classesin the first profile is lower. Therefore, the waiting component of the EF and BE traffic classes in the P1 profile is by 6.9 and 13% higher in comparison with the P2 profile since the transfer of these classes is completed faster. In case of the highest network load, the average packet delay of the AF traffic class is by 8.5% higher in the FWPBA_P1 then in the FWPBA_P2 where this class is less presented. As expected, the AF waiting component is equal to zero in the P1 profile because this class is the dominant one in almost every cycle. In the second profile, AF and BE traffic classes share the same amount of bandwidth but the AF class most frequently dictates the cycle duration. The reason for this behavior lies in the fact that bandwidth-intensive multimedia applications, which are characterized by large packets, are found within the AF traffic class. Consequently, in case of the implementation of the P2 profile, the waiting component for this class is almost annulated, just as it was the case in the first profile. The difference in waiting times of AF and BE traffic classes in the P2 profile is in the range of 15 µs since both classes occupy the same amount of bandwidth (40%).

B. *Unused bandwidth*

Simulation results presented in Fig. 7.3 further confirm the results of the analysis of the unused bandwidth in the FWPBA model which were conducted within Sect. 6.3. Consistent with the presented analysis, EF traffic has the worst characteristics since it is the least present in both profiles and is followed by BE traffic. BE traffic occupies the same amount of bandwidth as the medium priority class, but in terms of the waiting time, the actual requested bandwidth for AF class is higher since this class includes bandwidth-hungry applications. On the other hand, AF traffic is the most dominant regardless of the implemented profile so the bandwidth allocation is based on the bandwidth requirement of this class. Consequently, as we have expected, the AF characteristics are the best. In case of the implementation of the P1 profile, in which EF and BE traffic classes are less present in comparison

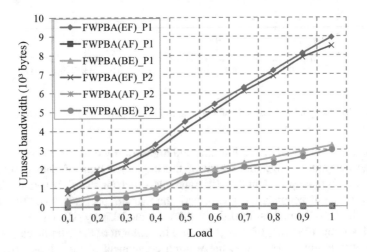

Fig. 7.3 Unused bandwidth per traffic class in FWPBA model

with the P2 profile, the unused bandwidth components at highest load are by 5 and 7% higher than the unused bandwidth component measured in the P2 profile.

C. *Jitter*

The packet delay variation or jitter is the next important QoS parameter for the assessment of network performance. For the successful transmission of EF traffic class (highest priority class for voice transmission) through the network, it is vital that the value of jitter be as low as possible. The jitter is represented by the packet delay variation of two consecutively departed EF packets from the same ONU in the same transmission window [48]:

$$J_i = D_i - D_{i-1} \tag{7.1}$$

where J_i is the ith delay jitter within the window and D_i is the ith packet delay within the window.

Figure 7.4 shows the probability density function (pdf) of the EF service packet delay at full loading scenario for both models. It is shown that the EF delay sequence presents dispersion with enough number of data points in a tail prior to 2 ms for the P2 profile and centralization with all data points condensed before 1.8 ms for the P1 profile. This further confirms that the model can provide the adequate QoS for a highest priority traffic class regardless of the implemented traffic traffic profile.

D. *Packet loss rate*

As illustrated in Fig. 7.5, the percentage of lost packets in the system is very small for both traffic profiles and this is the result of the increased efficiency of the OLT which originates from the use of a simple mathematical model and a fixed

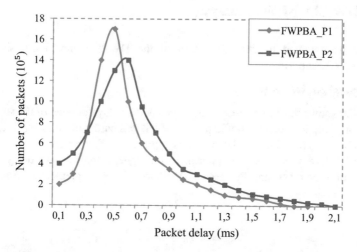

Fig. 7.4 Comparison of jitter performance in FWPBA model

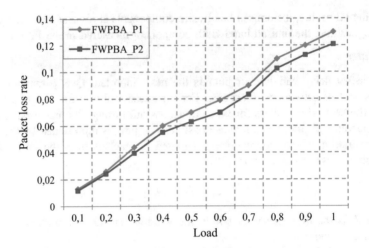

Fig. 7.5 Comparison of packet loss rates in FWPBA model

allocation of wavelengths. Packet loss in the models mainly comes from the fact that packet fragmentation is not allowed. Therefore, this large packet that does not fit the currently granted window will have to be postponed to the next granted window. At the highest load, the measured packet loss in the P1 profile is by 7.7% higher than packet losses measured in case of the implementation of the P2 profile. Such a result is to be anticipated since AF traffic class is the most present traffic class within the first profile and such packets can be very large (video streaming, multimedia transmission).

7.3 DWPBA Model Analysis

The results obtained by the simulation of the DWPBA model are shown in Figs. 7.6, 7.7, and 7.8.

A. *Average packet delay*

Figure 7.6 shows the comparison of average packet delaysfor each of the supported classes of traffic for different traffic profiles. As it was case with the FWPBA model, the obtained results confirm the previously presented theoretical analysis. Given that the EF traffic class is least present in the system, the average delay characteristics of this class are the best regardless of the implemented profile. At

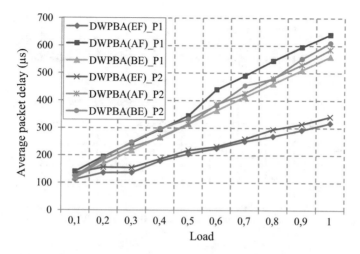

Fig. 7.6 Average packet delay in DWPBA model

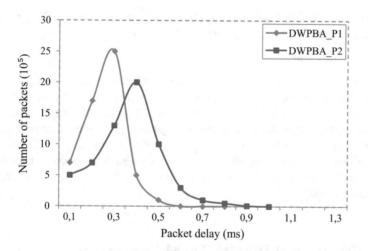

Fig. 7.7 Comparison of jitter performance in DWPBA model

maximum load, the average packet delay of the EF traffic class measured in case of the implementation of the P1 profile is up to 6.8% better than the delay measured in case of the implementation of the P2 profile and this could be explained by the fact that the BE traffic class is less present in the first profile than in the second profile.

B. *Jitter*

Jitter characteristics are presented in Fig. 7.7. As illustrated, the EF delay sequence presents dispersion with enough number of data points in a tail prior to 0.95 ms for the P2 traffic profile and is reduced to 0.6 ms for the P1 traffic profile. This further

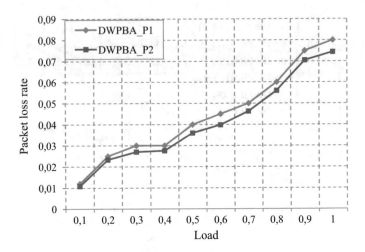

Fig. 7.8 Comparison of packet loss rates in DWPBA model

confirms that the DWPBA model is able to provide an excellent EF jitter performance regardless of the implemented traffic profile.

C. *Packet loss rate*

The characteristics of the packet loss rates for different traffic profiles are presented
in Fig. 7.8. At the highest load, the measured packet loss in the P1 profile is by
7% higher than the packet loss measured in the P2 profile. Once again, this is to be
expected as the AF traffic class is the most present class in the first profile. Given
the fact that these packets can be very large (video streaming, multimedia transmission) and that packet fragmentation is not allowed, large packets that do not fit
the currently granted window will be postponed to the next granted window and
consequently lost.

7.4   FWPBA and DWPBA Model Comparison

In this section, we present the results of the comparison of the proposed models in
case of the implementation of the first traffic profile which describes the current
situation in the access network. The presented results are obtained from the simulation of the WDM EPONwith 16 ONUs.

A. *Average packet delay*

Firstly, we analyze the results for average and maximum packet delays (Figs. 7.9
and 7.10). Simulation results show significant improvement in delay characteristics,
especially for EF traffic in case of the implementation of the DWPBA model.
Compared with the EF characteristics in the FWPBA model, the EF traffic class

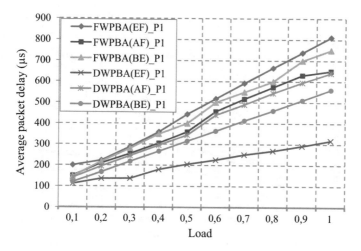

Fig. 7.9 Comparison of average packet delays in FWPBA and DWPBA models

now has better characteristics by up to 61% in terms of the average delay and by 64.8% in terms of the maximum packet delay. At the same time, BE characteristics show better results by 25 and 29.4% in terms of the average and maximum packet delay, respectively, and this is so because the waiting time does not exist in the system any longer. Since the waiting time has the highest value for EF traffic, this class now experiences the smallest delay. At the same time, the DWPBA model does not degrade the performance of medium priority traffic class. Moreover, compared with the characteristics of this class in the FWPBA model, this model improves the characteristics of both average and maximum packet delay of the AF traffic class by 1.5 and 2.7%, respectively.

B. *Jitter*

Better performance of the DWPBA model is further confirmed by the analysis of the EF packet delay variation at a full loading scenario and EF, AF and BE average queue occupancy (Figs. 7.11 and 7.12). Figure 7.11 shows the probability density function of the EF service packet delay at the highest network load for both models. EF delay sequence presents dispersion with enough number of data points in a tail prior to 1.8 ms for FWPBA, and centralization with all data points condensed before 0.6 ms for the DWPBA model. Accordingly, the DWPBA model presents the optimal EF delay variation. The average queue occupancy has significantly decreased in case of the implementation of the DWPBA (by 47.4% for EF traffic). As it was the case with packet delays, the transmission of AF traffic has not been degraded with the implementation of the DWPBA. Moreover, simulation results show that queue occupancy has decreased by 3% due to increased system efficiency.

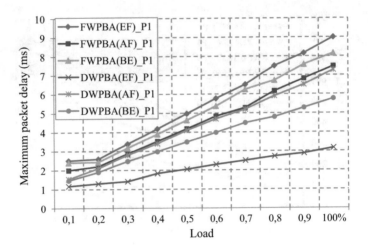

Fig, 7.10 Comparison of maximum packet delays in FWPBA and DWPBA models

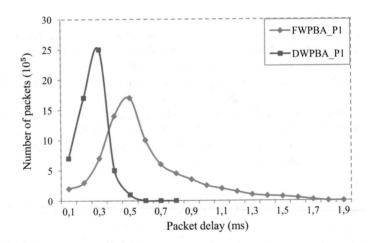

Fig. 7.11 Comparison of jitter performance in FWPBA and DWPBA models

C. *Packet loss*

Next, we compare the packet loss characteristics as well as the network throughput characteristics. As we have previously explained, in the FWPBA model, the additional packet loss could be induced by the presence of waiting time as well as by the fact that lower loaded wavelengths are likely to be underutilized while waiting for the most loaded wavelength to finish the transmission. In the absence of

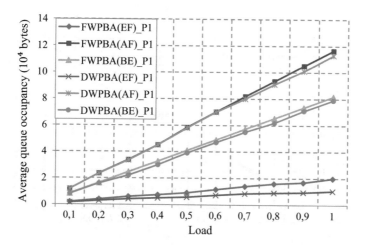

Fig. 7.12 Comparison of average queue occupancy in FWPBA and DWPBA models

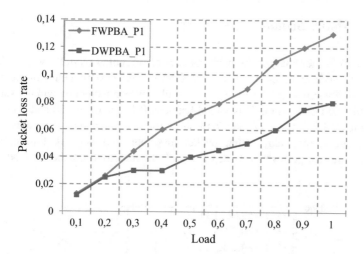

Fig. 7.13 Comparison of packet loss rates in FWPBA and DWPBA models

waiting time, the DWPBA model has better packet loss characteristics by up to 1.6 times (Fig. 7.13). This further induces better network throughput characteristics in case of the implementation of the DWPBA model. Network throughput analysis shows improvement within the framework of the DWPBA model (92%), compared with the FWPBA model (87%) (Fig. 7.14). This further confirms that the DWPBA model has better performance in comparison with the FWPBA model.

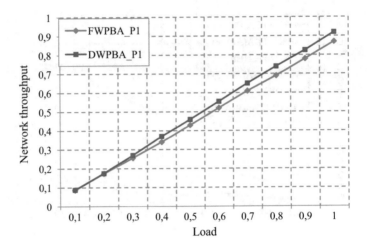

Fig. 7.14 Comparison of network throughput in FWPBA and DWPBA models

7.4.1   Model Scalability Analysis

In order to further analyze the performances of presented models, we simulate the access network with 64 implemented ONUs, i.e., system behavior in case of the maximum supported load [95]. The simulation parameters remain the same and the traffic load of each ONU is varied between 0.1 and 1 (i.e., 10 and 100 Mbps).

Figures 7.15, 7.16, 7.17, and 7.18 show the characteristics of models, namely the average traffic delay, jitter, packet loss rate, and network throughput. As expected, system performances degrade with the increased number of ONUs since

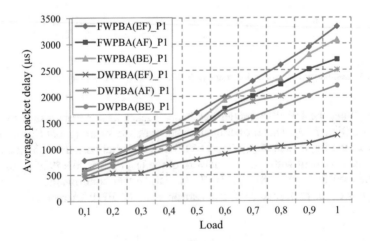

Fig. 7.15 Comparison of average packet delays in FWPBA and DWPBA models for 64 ONUs

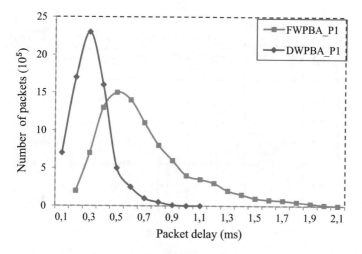

Fig. 7.16 Comparison of jitter performance in FWPBA and DWPBA models for 64 ONUs

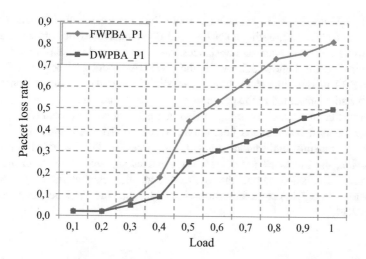

Fig. 7.17 Comparison of packet loss rates in FWPBA and DWPBA models for 64 ONUs

the network load now exceeds the maximum supported load. In case of the WDM EPON with 16 ONUs, the total highest load generated in the network is 16×100 Mbps = 1.6 Gbps, which is much smaller than the total available bandwidth of 4 Gbps (four wavelengths with throughput of 1 Gbps). This clearly shows that all wavelengths are not fully utilized, and hence, smaller delays (see Figs. 7.9, 7.10, 7.11, 7.12, 7.13, and 7.14) are experienced. On the contrary, in the system with 64 ONUs, the total highest load generated in the network is

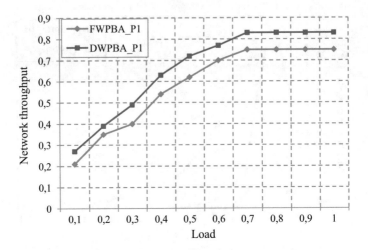

Fig. 7.18 Comparison of network throughput in FWPBA and DWPBA models for 64 ONUs

64×100 Mbps = 6.4 Gbps, which is more than the total available bandwidth of 4 Gbps, and therefore, both packet loss and delay characteristics are degraded.

However, even in such circumstances, the average packet delay in case of the maximum load is bellow 3.5 ms for the FWPBA model and bellow 2.5 ms for the DWPBA model. Jitter characteristics are slightly degraded in comparison with the characteristics presented in the previous section (compare Figs. 7.11 and 7.16). The characteristics of packet loss rate are also degraded since the probability of packet loss is increased but remains below 1%, which confirms the system's ability to efficiently transmit multimedia content regardless of the increased number of ONUs, i.e., the increased input load. As expected, in accordance with the afore-mentioned discussion about the system load, the characteristics of network throughput (Fig. 7.18) come into saturation faster than the characteristics presented in Fig. 7.14.

7.4.2   Comparison of Results Among Different DWBA Models

In order to further analyze the effectiveness and performance of the FWPBA and DWPBA, we compare the obtained results with the results presented in [89, 91, 92] since these papers discuss wavelength and bandwidth allocation in WDM EPONs. For the sake of comparison, we make use of the results obtained by the imple-mentation of both traffic profiles depending on the profile that has been imple-mented in the aforementioned papers. It should be emphasized that the comparison of simulation results is not completely accurate since simulation parameters are not entirely the same (however small, deviations still exist).

First, we compare the performance of the presented models with the three-channel Diffserv WDM IPACT-ST model presented in [89]. At the highest network load, the average packet delay of the EF traffic class decreases for one and two orders of magnitude in FWPBA and DWPBA models, respectively, compared with the WDM IPACT-ST, and for three orders of magnitude for both the AF and BE traffic classes. Besides that, the packet loss rate decreases for two orders of magnitude compared with the results given in [89].

Authors in [91] present three different DWBA models for hybrid TDM/WDM network that supports the extension of the MPCP protocol. However, the authors do not discuss the QOS implementation for either of these models. The presented solution has gained a lot of attention recently since it suggests a possible simple migration from the TDM-PON to the TDM/WDM-PON. For the purpose of our analysis, we compare the FWPBA and DWPBA model performances only with the DWBA-2 (FE) because this model supports the same number of wavelengths as our models. Since the DWBA-2 (FE) model does not support QoS and traffic classes, we used the sum of the average packet delays in the FWPBA and DWPBA models for the purpose of comparison. At the highest network load, the sum of average packet delays of all three supported traffic classes in FWPBA and DWPBA models is two orders of magnitude smaller than the average packet delay presented in [91].

In [92], authors discuss a WDM EPON system that supports traffic transmission on two defined wavelengths and QoS implementation. Since the authors discuss the implementation of different traffic profiles, the presented comparison is not completely precise. Furthermore, the authors present three models for wavelength and bandwidth allocation, namely the QoS-DBA-1, QoS-DBA-2, and QoS-DBA-3 model. The average packet delay of the AF and BE traffic classes in both models is for one to three (depending on the implemented traffic profile) orders of magnitude smaller compared with the QoS-DBA models. Packet loss rate decreases for one order of magnitude compared with the results given in [92]. Consequently, network throughput also decreases by 8–42.5% depending on the DWBA model and the implemented traffic profile. Besides that, at the highest network load, the EF delay sequence presents dispersion with enough data points in a tail prior to 6 ms for the QoS-DBA-1 model compared with 2.1 and 0.9 ms for the FWPBA and DWPBA models, respectively, which further confirms the superior performance of our models.

7.4.3 Comparison with a Single-Channel EPON

Finally, in order to validate the introduction of the WDM technology in classic EPONs, we compare the results obtained from the simulation of FWPBA and DWPBA models with the performance of a single-channel EPON system, i.e., the HG(PBS) model described in the fourth chapter.

The traffic profile implemented in the HG(PBS) model is as follows: 20% of the total generated traffic is allocated for narrowband EF service, whereas the remaining

80% is equally distributed between the and BE services [48]. Accordingly, for the sake of comparison with this model, we use the results obtained by the FWPBA and DWPBA testing in case of the P2 profile implementation. For the same reason we simulate the HG(PBS) model with 64 ONUs instead of the system with 32 ONUs presented in [48].

The presented simulation results of both WDM EPON models show a significant improvement in the overall system performance in comparison with the HG(PBS) algorithm, Figs. 7.19, 7.20, 7.21, 7.22, and 7.23. The FWPBA simulation results show improvement in the average EF (5.73%) and considerable improvement in the

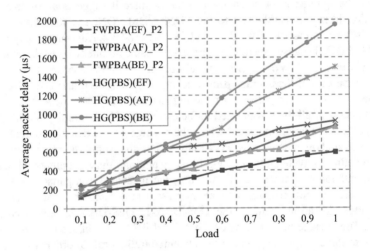

Fig. 7.19 Comparison of average packet delays in FWPBA and HG(PBS) models

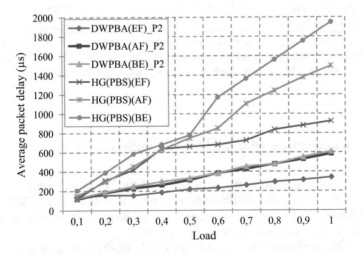

Fig. 7.20 Comparison of average packet delays in DWPBA and HG(PBS) models

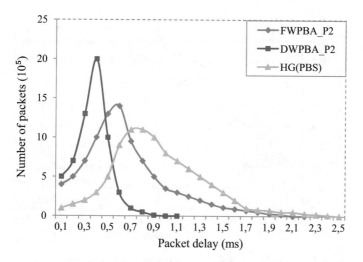

Fig. 7.21 Comparison of jitter performance in FWPBA, DWPBA and HG(PBS) models

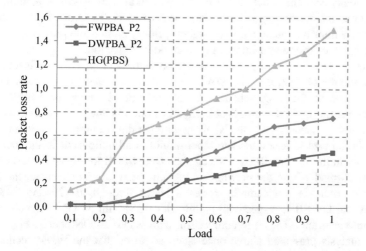

Fig. 7.22 Comparison of packet loss rates in FWPBA, DWPBA, and HG(PBS) models

average AF (2.5 times) and BE (2.2 times) packet delays at the load of one
(Fig. 7.19). As expected, the simulation results for the DWPBA model, as a more
efficient multichannel model, show considerable improvement in the average packet
delay of all three traffic classes at the load of one (2.7 times for EF, 2.6 times for
AF, and 3.2 times for BE), Fig. 7.20.

The jitter analysis is shown in Fig. 7.21. As presented, the EF delay sequence
presents a dispersion with enough number of data points in a tail until 2.1 ms for
FWPBA with P2 profile, reduced to 0.93 ms for DWPBA, and a centralization with

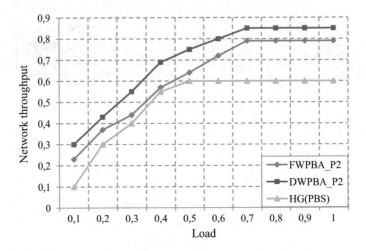

Fig. 7.23 Comparison of network throughput in DWPBA and HG(PBS) models

all data points condensed before 2.5 ms for HG(PBS) with 64 ONU. Although the last scheme maintains a good EF jitter performance, the multichannel models significantly improve transmission of highest priority class while providing fairness to other types of traffic and improving bandwidth utilization.

The characteristics of the packet loss rate presented in Fig. 7.22 confirm that the implementation of FWPBA and DWPBA significantly decreases packet loss in comparison with the single-channel EPON system (2 times for FWPBA and 3.2 times for the DWPBA model at the load of one). Namely, the increased input load (64 ONUs instead of 32 ONUs) along with the limited buffer sizes (there is not enough buffer size to accommodate input traffic) in all units leads to the increased packet loss in the system. The enhanced characteristics of multichannel models are finally confirmed within the network throughput analyses. As expected, the intro-duction of multiple wavelengths raises network throughput (79 and 85% for FWPBA and DWPBA, respectively) while HG(PBS) with 64 ONUs much sooner comes into saturation (60%) in comparison with WDM EPONmodels, Fig. 7.23.

The analysis presented above once again confirms that the WDM technology will certainly become a necessity for the realization of next generation access networks.

7.5 The DWPBA-FS Model Analysis

The DWPBA-FS model is tested using a WDM EPON network model developed in MATLAB, using Simulink packet and the same simulation parameters as in the case of FWPBA and DWPBA models. In simulations, we consider a WDM EPONconsisting of 64 WDM ONUs that support traffic transmission on four

different wavelengths in each station and the speed of each wavelength amounts to
1 Gbps.

As we have previously explained, traffic schemes in next generation networks
(NGNs) have been recently changed, and the AF traffic class has now become
predominant [1]. Consistently, we adopt the P1 traffic profile in order to test and
analyze the FWPBA_FS model, i.e., 15% of the total generated traffic is considered
to be high priority, 50% is considered to be medium priority multimedia traffic of
the AF class, and the remaining 35% is considered to be low-priority BE traffic.
Weighted proportions used in the WFQ mechanism for scheduling of the AF
subclasses are 0.2, 0.1, 0.5, and 0.2 for Q4, Q3, Q2, and Q1, respectively. The
communication quantities used for the comparison are the following: average
packet delay, jitter, throughput rate, and packet loss rate. In accordance with the fact
that the DWPBA_FS is directly developed from the DWPBA model, the obtained
results are analyzed and compared with the DWPBA model characteristics
described and analyzed in the preceding sections.

A. *Average packet delay*

We first analyze the average packet delay per wavelength, Fig. 7.24. We analyze
the behavior of the EF and BE traffic classes first of all, since the elimination of the
synchronization wavelength could potentially degrade their transmission characteristics. The simulation results show that the average packet delays for the EF and
BE classes increases (a negligible, slight degradation of characteristics) by 1.7 and
2.1%, respectively, in comparison with the DWPBA model. At the same time, the
transmission of the AF traffic class on the two defined wavelengths results with
delay (measured per wavelength) below 1.5 ms decreases by 42.6% in comparison
with its predecessor, i.e., the DWPBA model tested under the same conditions with
64 ONUs. As expected, the simulation results follow the load distribution between

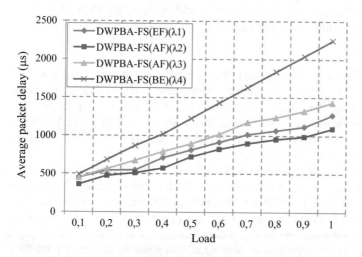

Fig. 7.24 Average packet delay per wavelength in DWPBA-FS model

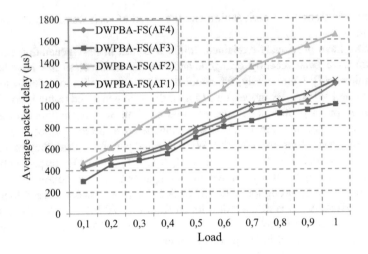

Fig. 7.25 Average packet delay of AF subclasses in DWPBA-FS model

the subclasses, and consequently, the AF3 subclass, which is the least represented in the system, has the best characteristics of delay, while the most presented AF2 subclass has the largest delay, Fig. 7.25.

The presented results confirm the excellent performance of the DWPBA-FS model because the average packet delays of all traffic classes are less than 2.3 ms, which enables QoS provision and the efficient delivery of not only the multimedia traffic, which was the primary goal, but also the low-priority traffic class transmission as well. Since the analyses conducted within the previous section confirmed that FWPBA and DWPBA models provide better performances for all tested parameters in comparison with HG(PBS) (single-channel model); in further analyses, we compare only multichannel models with the DWPBA-FS model in order to validate the new approach and the introduction of AF traffic subclasses.

B. *Jitter*

The variation of the EF packet delay, i.e., jitter is the next QoS parameter that must be analyzed in order to confirm the performances of the DWPBA-FS model. Jitter is essential for the successful transmission of highest priority traffic. As we have explained in the previous sections, jitter is represented by the packet delay variation of two consecutively departed EF packets from the same ONU in the same transmission window [48, 51–52].

Figure 7.26 shows the probability density function (pdf) of the EF service packet delay at full loading scenario for the FWPBA, DWPBA, and DWPBA-FS models in case of the P1 profile implementation. The presented results confirm that the DWPBA-FS model is able to provide an excellent EF jitter and at the same time enhance the transmission of medium priority traffic. The EF delay sequence presents dispersion with enough number of data points in a tail until 1 ms for DWPBA, and 1.2 ms for the DWPBA-FS (a negligible degradation of only 0.2 ms) and

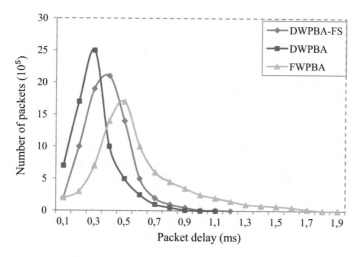

Fig. 7.26 Comparison of jitter performances

1.9 ms for the FWPBA model. As it was the case with average packet delay characteristics, the elimination of the synchronization wavelength degrades the jitter characteristics, but this degradation is negligible and does not significantly affect the efficiency of the priority traffic class transmission.

C. *Throughput and packet loss rate*

Besides the delay analyses, we further analyze the throughput and packet loss rate as essential QoS parameters for the assessment of network performance.

Figure 7.27 shows the comparison of throughput for the AF traffic class (one wavelength in FWPBA and DWPBA, two wavelengths in the DWPBA-FS model). The FWPBA (AF) and DWPBA (AF) characteristics relatively quickly come into saturation (at the input load below 0.5), while the DWPBA-FS model distributes load on two wavelengths and at the highest load, the AF throughput is doubled. Moreover, the synchronization wavelength of FWPBA and DWPBA is used for data transmission in DWPBA-FS, and as a result, the overall network throughput is increased. As shown in Fig. 7.28, the DWPBA-FS model improves the overall throughput rate in comparison with DWPBA and FWPBA (i.e., at full loading, DWPBA-FS throughput rate is 89%, DWPBA throughput rate is 83%, and FWPBA throughput is 75%). The same behavior applies to the packet loss rate performance in all schemes, Fig. 7.29. Packet loss in the system originates mainly from the fact that packet fragmentation is not allowed and therefore large packets which do not fit the currently granted window will have to be postponed and consequently lost. At

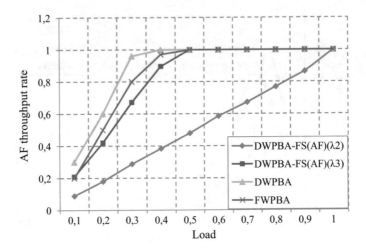

Fig. 7.27 Comparison of AF throughput rate

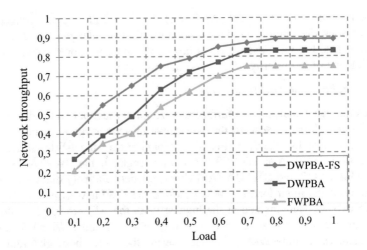

Fig. 7.28 Comparison of network throughput

the same time, the AF traffic class is the most present traffic class and is often characterized with multimedia that can be very large and therefore needs fragmentation. Namely, different multimedia applications that involve streaming and real-time transmission are classified as medium priority traffic class [1]. With the

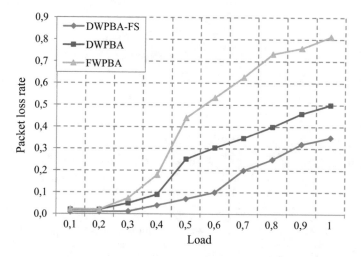

Fig. 7.29 Comparison of packet loss rates

implementation of the AF subclasses, the WFQ mechanism, and the two wave-lengths for the transmission of the AF subclasses (the probability of large packets being transmitted immediately is increased), packet loss is significantly decreased. At the highest network load, packet loss in the DWPBA-FS is decreased by 38% in comparison with the DWPBA, and 2.3 times in comparison with the FWPBA model that further proves the superiority of the DWPBA-FS model.

Chapter 8
Conclusion

Since the next wave in the development of next generation networks has already begun, service providers are investing in the development of new solutions in order to foster communications, expand the service portfolio, and create a greater value through various services and multimedia-based applications.

As we have discussed in the introductory section, the evolution of the first mile is, for the most part, the result of the technology breakthroughs in telecommunications, such as WDM, which made bandwidth cheap and super-fast. As a result of this evolution, a practical and ubiquitous worldwide Internet is enabled, as well as broadband access networks along with the introduction of the triple-play networks, Fig. 8.1. Most IP traffic growth results from growth in Internet traffic, compared to managed IP traffic. Of the total of 80.5 exabytes, 60 are due to fixed Internet and 6 are due to mobile Internet. Fixed and mobile Internet traffic, in turn, is propelled by video. As we have outlined in some of our previous forecasts, the sum of all forms of IP video (Internet video, IP VoD, video files exchanged through file sharing, video-streamed gaming, and video conferencing) will ultimately amount for 90% of the total IP traffic.

The adoption of the WDM technology in the core, as well as metropolitan networks, and the introduction of the gigabit Ethernet in the LANs of most organizations could successfully resolve the bandwidth issue in these networks. However, the anticipated bandwidth growth at the same time demands the increase in the bandwidth available in the access network along with the improved QoS parameters. In such circumstances, the further development of access networks is perceived as a key factor for the further development of the NGN and Internet itself.

According to the research conducted in [103], the future will bring the following:

- The number of households generating over 1 terabyte per month of Internet traffic will reach 1 million by the end of 2012;
- Annual global IP traffic will reach the zettabyte threshold (966 exabytes or nearly 1 zettabyte) by the end of 2015;

© Academic Mind and Springer International Publishing AG 2017   217
M. Radivojević and P. Matavulj, *The Emerging WDM EPON*,
DOI 10.1007/978-3-319-54224-9_8

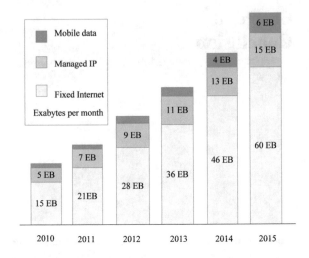

Fig. 8.1 Global IP traffic forecast. (http://www.cisco.com/en/US/solutions/collateral/ns341/ns525/ns537/ns705/ns827/VNI_HyperconnectivityWP.html)

- In 2015, the gigabyte equivalent of all movies ever made will cross global IP networks every 5 min. Global IP networks will deliver 7.3 petabytes every 5 min in 2015;
- In 2015, there will be 6 million Internet households worldwide generating over a terabyte per month in Internet traffic, up from just a few hundred thousand in 2010. There will be over 20 million households generating half a terabyte per month in 2015;
- The number of devices connected to IP networks will be twice as high as the global population in 2015;
- A growing amount of Internet traffic is originated with non-PC devices. In 2010, only 3% of Internet traffic was originated with non-PC devices, but by 2015, the non-PC share of Internet traffic will grow to 15%;
- Global Internet video traffic surpassed global peer-to-peer (P2P) traffic. Internet video will reach 50% by the end of 2012;
- It would take over 5 years to watch the amount of video that will cross global IP networks every second in 2015. Every second, 1 million minutes of video content will cross the network in 2015;
- Internet video now makes 40% of consumer Internet traffic, and will reach 62% by the end of 2015, not including the amount of video exchanged through P2P file sharing. The sum of all forms of video (TV, VoD, Internet, and P2P) will continue to amount to approximately 90% of the global consumer traffic by 2015;
- Video-on-demand traffic will triple by 2015. The amount of VoD traffic in 2015 will be equivalent to 3 billion DVDs per month;
- High-definition VoD will surpass standard definition by the end of 2011. By 2015, high-definition Internet video will comprise 77% of VoD.

It is obvious that various new applications demand the increase in the bandwidth available in the access networks along with the ultimately improved QoS parameters. It means that the further development of the access networks must be encouraged for the sake of the further development of the NGN and Internet itself. Today, such a development is tightly connected with the development of the optical access technologies as the cornerstone for the development of the truly broadband access network.

In recent years, most of the FTTH deployments have been based on industry's standard optical access networks, such as EPON and GPON. The success of these deployments has led to significant innovations in both system architecture and in terms of components which are used to build these systems. Hence, the next generation of PONs will inevitably be far more advanced than what is typically deployed today. At the forefront of PON development, there have been two separate approaches that appear to respond to all of the demands of the next generation systems: 10 Gbps PON (10 G EPON or 10 G GPON in the relevant literature) and WDM PON. Each approach has its own advantages and its own drawbacks, but recently, progress has been significantly accelerated by means of both new technologies.

This book presents the emerging WDM EPONs as the potential optimal solution for the realization of the broadband access infrastructure.

The main advantages of WDM EPONs are as follows:

- The optical distribution plant is still passive and therefore has the same low-maintenance and high-reliability properties;
- Excellent privacy;
- P2P connections between the OLT and ONUs are realized in the wavelength domain. There is no P2MP media access control required, and consequently the MAC layer is substantially simplified;
- There is no distance limitation imposed by the ranging and DBA protocols;
- Easy 'pay as you grow' upgrade. Each wavelength in a WDM PON can run at a different speed as well as with different protocols. The individual user pays for an upgrade.

At the same time, one of the main challenges of the WDM PON includes the high costs of WDM components. However, since the costs of components have decreased tremendously in recent years, this has made WDM PONs economically more viable. Today, the main goal of ISPs is to migrate gradually from extant TDM PON to the WDM PON facilitating operators to replace users and network equipment. Such an upgrade is also supported and driven by the IEEE and ITU-T through the development of new standards.

In this book, we have presented the development and the characteristics of the EPON system, as well as the protocols and services that are used in access networks. Special emphasis has been placed on the techniques, and support for QoS implementation in the access network since the QoS support has become necessary for the successful realization and further development of future communication networks. Moreover, we have presented the WDM EPON architecture with the aim of solving bandwidth limitations of the current single-channel EPONs. Two novel dynamic wavelength and bandwidth allocation (DWBA) models, namely the fixed wavelength priority bandwidth allocation (FWPBA) and the dynamic wavelength priority bandwidth allocation (DWPBA) for hybrid TDM/WDM EPON have been proposed as well. These models use the existing MPCP protocol defined by the IEEE 802.3ah working group and are very flexible with respect to WDM architectures that may evolve in the OLT and ONUs. Accordingly, they could be easily upgraded on a per need basis in terms of the supported wavelengths and the number of ONUs. Additionally, both models could be easily implemented in the existing EPON systems in a timely and appropriate manner.

Furthermore, we have presented a new approach for the analysis of QoS in WDM EPONs in which we use wavelength assignment per service class and not per ONU, as it has often been suggested by the common approach in the relevant literature. In this way, we avoid the need for the implementation of additional complex algorithms for supporting QoS which further reduces system cost and increases the overall system efficiency.

The theoretical analysis and the comparison of the advantages and disadvantages of the aforementioned models have also been provided. Apart from the theoretical analysis, we have studied and evaluated the performance of both models through the detailed simulation experiments. System characteristics have been tested through extensive simulations by means of an original network model developed in MATLAB which incorporates all key parameters of the real network environment, such as queuing delay, transmission delay, and packet processing delay, round-trip time, and framing overhead. The simulation results confirm the excellent performance of the presented models in terms of average packet delay, jitter, packet loss, and throughput, as follows:

- In comparison with single-channel HG(PBS) model, the FWPBA model simulation results show improvement in the average EF (5.73%), AF (2.5 times), and BE (2.2 times) packet delays. The simulation results for DWPBA model, as a more efficient multichannel model, show a considerable improvement in the average packet delay of all three traffic classes (2.7 times for EF, 2.6 times for AF, and 3.2 times for BE). The jitter analysis also shows improvement: The EF delay sequence is condensed before 2.1 ms for FWPBA and reduced to 0.9 ms in DWPBA, while in the HG(PBS) model, EF delay sequence presents a dispersion with enough number of data points in a tail until 2.5 ms. Moreover, the characteristics of the packet loss rate confirm that the implementation of FWPBA and DWPBA significantly decreases packet loss in comparison with the single-channel EPON system (2 times for FWPBA and 3.2 times for the DWPBA model). Consequently, the introduction of multiple wavelengths improves network throughput (79 and 85% for FWPBA and DWPBA, respectively) while HG(PBS) with 64 ONUs much sooner comes into saturation (60%);

- In comparison with the three-channel Diffserv WDM IPACT-ST model presented in [89], the average packet delay of the EF traffic class decreases for one and two orders of magnitude in the FWPBA and DWPBA models, respectively, compared with a three-channel WDM IPACT-ST [89], and for three orders of magnitude for both the AF and BE traffic classes. Besides that, packet loss rate decreases for two orders of magnitude compared with the results given in [89];
- In comparison with the DWBA-2 (FE) model (without QoS support) presented in [91], the sum of average packet delays of all three supported traffic classes in the FWPBA and DWPBA models is two orders of magnitude smaller than the average packet delay of DWPBA-2 (FE);
- In comparison with the three QoS-DBA models for wavelength and bandwidth allocation presented in [92], the average packet delay of the AF and BE traffic classes in the FWPBA and DWPBA is for one to three (depending on the implemented traffic profile) orders of magnitude smaller. At the highest network load, the EF delay sequence presents dispersion with enough data points in a tail prior to 6 ms for the QoS-DBA-1 model compared with 2.1 and 0.9 ms for the FWPBA and DWPBA models, respectively. Packet loss rate decreases for one order of magnitude compared with the results given in [92]. Consequently, network throughput also increases by 8–42.5% depending on the DWBA model and the implemented traffic profile.

Based on the obtained simulation results and previous analyses, the FWPBA and DWPBA models appear to be superior in comparison with other DWBA models which have been proposed so far. However, we have suggested and analyzed the further development of the DWPBA model called the DWPBA-FS which includes a finer segregation of multimedia-based traffic, i.e., a model which introduces new traffic classes in accordance with the Diffserv framework.

The simulation results show that the transmission of the AF traffic class on the two defined wavelengths results in delay (measured per wavelength) below 1.5 ms (decreases by 42.6 and 39% in comparison with the DWPBA and FWPBA models, respectively). The EF delay sequence presents dispersion with enough number of data points in a tail until 1 ms for DWPBA, and 1.2 ms for the DWPBA-FS (a negligible degradation of only 0.2 ms), and 1.9 ms for the FWPBA model. The DWPBA-FS model also improves the overall throughput rate in comparison with DWPBA and FWPBA (i.e., at full loading scenario, the DWPBA-FS throughput rate is 89%, the DWPBA throughput rate is 83%, and the FWPBA throughput is 75%). Moreover, at the highest network load, packet loss in the DWPBA-FS is decreased by 38% in comparison with the DWPBA, and 2.3 times in comparison with the FWPBA model.

Simulation results confirm the superiority of the DWPBA-FS model and with the introduction of new subclasses, such as WDM EPONs are capable of efficiently managing different video-based applications that will most likely be prevalent in future networks. Given the stated reasons, we believe that the DWPBA-FS model may become an eligible candidate for wavelength and bandwidth allocation in next generation EPON networks. Moreover, since the network model incorporates all key parameters of multimedia ISP network, we are convinced that the implementation of a hardware for the solution we here propose would enable network operation in a manner which could be easily foreseen and would therefore not involve any complexities related to the implementation of additional algorithms which are commonly found within other solutions which have been proposed so far.

References

1. F.J. Hens, J.M. Caballero, *Triple Play: Building the Converged Network for IP, VoIP and IPTV*, vol. 1 (Wiley, West Sussex, UK, 2008), pp. 1–28
2. OECD Telecom and Internet Reports, National broadband plans [pdf] (2011). http://www.oecd.org/dataoecd/22/41/48459395.pdf
3. Government of Republic Serbia, Strategy for development of broadband access in Republic Serbia by 2012 [pdf, Serbian version] (2009). http://www.mtid.gov.rs/wp-content/uploads/Dokumenti/Strategije_akcioni_planovi/Strategija_i_akcioni_plan_razvoj_sirokopojasnog_pristupa.pdf
4. S. Palm, ITU-T xDSL Standards [pdf] (2004). http://www.itu.int/ITU-T/worksem/asna/presentations/Session_6/asna_0604_s6_p4_palm.pdf
5. Broadband forum, DSL technology evolution [pdf] (2009). http://www.broadband-forum.org/downloads/AboutDSL.pdf
6. ITU Recommendation ITU-T G992.1, Asymmetric digital subscriber line (ADSL) transceivers [pdf] (1999). http://www.itu.int/rec/recommendation.asp?type=folders&lang=e&parent=T-REC-G.992.1
7. ITU Recommendation ITU-T G992.2, Splitterless asymmetric digital subscriber line (ADSL) transceivers [pdf] (1999). http://www.itu.int/rec/recommendation.asp?type=folders&lang=e&parent=T-REC-G.992.2
8. ITU Recommendation ITU-T G992.5. (2009). Asymmetric digital subscriber line 2 transceivers (ADSL2)—Extended bandwidth ADSL2 (ADSL2plus) [pdf]. http://www.itu.int/rec/recommendation.asp?type=folders&lang=e&parent=T-REC-G.992.5
9. ITU-T publication ITU-T G.995.1, Overview of digital subscriber line (DSL) Recommendations [pdf] (2001). http://www.itu.int/rec/recommendation.asp?type=folders&lang=e&parent=T-REC-G.995.1
10. Cable Europe Labs, Cable network handbook [pdf] (2009). http://www.cableeurope.eu/index.php?page=technology
11. CableLabs, DOCSIS specification—DOCSIS 3.0 Interface [pdf] (2011). http://www.cablelabs.com/cablemodem/specifications
12. J. Stenger, Broadband power line tutorial [pdf] (2009). http://www.wave-report.com
13. FTTH Council, Definition of terms [pdf] (2009). http://www2.ftthcouncil.eu/documents/studies/FTTH-DefinitionsRevision_January_2009.pdf
14. IEEE 802.11, Local and metropolitan area networks (2007). http://standards.ieee.org/getieee802/download/802.11-2007.pdf
15. IEEE 802.16, Air interface for broadband wireless access systems (2009). http://standards.ieee.org/getieee802/download/802.16-2009.pdf
16. S. Shepard, *WiMax Crash Course*, vol. 1 (The McGraw-Hill Companies, New York, NY, 2006), pp. 1–44 (0-07226-307-5)
17. M.S. Kuran, T. Tugcu, Comput. Netw. **51**, 3013–3046 (2007)

18. J.C. Crimi, Next generation network services [pdf] (2009). http://www.mobilein.com/NGN_Svcs_WP.pdf
19. F. Effenberger, G. Cleary, O. Haran, G. Kramer, R. Ding Li, M. Oron, T. Pfeiffer, IEEE Commun. Mag. **1**, 17–25 (2007)
20. W. Odom, M.J. Cavanaugh, *Cisco QoS*, vol. 2 (Cisco Press, Indianapolis, IN, 2005), pp. 83–141 (1-58720-124-0)
21. M. Flannagan, R. Froom, K. Turek, *Quality of Service in Campus Networks*, vol. 1 (Cisco Press, Indianapolis, IN, 2003), pp. 22–94 (1-58705-120-6)
22. F. Baker, RFC 3393 [txt] (1995). http://www.ietf.org/rfc/rfc1812.txf7numberf393
23. R. Braden, RFC 1633 [txt] (1994). http://www.ietf.org/rfc/rfc1633.txt?number=1633
24. D. Durham, J. Boyle, R. Cohen, S. Herzog, R. Rajan, A. Sastry, RFC 2748 [txt] (2000). http://www.rfc-editor.org/rfc/rfc2748.txt
25. S. Herzog, J. Boyle, R. Cohen, D. Durham, R. Rajan, A. Sastry, RFC 2749 [txt] (2000). http://www.rfc-editor.org/rfc/rfc2749.txt
26. K. Chan, J. Seligson, D. Durham, S. Gai, K. McCloghrie, S. Herzog, F. Reichmeyer, R. Yavatkar, A. Smith, RFC 3084 [txt] (2001). http://www.rfc-editor.org/rfc/rfc3084.txt
27. K. Nichols, S. Blake, F. Baker, D. Blask, RFC 2474 [txt] (1998). http://www.rfc-editor.org/rfc/rfc2474.txt
28. B. Davie, A. Charny, J.C.R. Bennett, K. Benson, J.Y. Le Boudec, W. Courtney, S. Davari, V. Firoiu, D. Stiliadis, RFC 3246 [txt] (2002). http://www.rfc-editor.org/rfc/rfc3246.txt
29. J. Heinanen, F. Baker, W. Weiss, J. Wroclawski, RFC 2597 [txt] (1999). http://www.rfc-editor.org/rfc/rfc2597.txt
30. D. Grossman, RFC 3260 [txt] (2002). http://www.rfc-editor.org/rfc/rfc3260.txt
31. N. Ek, IEEE 802.1 P,Q—QoS on the MAC level [html] (1999). http://www.tml.tkk.fi/Opinnot/Tik-110.551/1999/papers/08IEEE802.1QosInMAC/qos.html
32. P.D. Matavulj, J. Radunovic, J. Infrared Millim. Terahertz Waves **22**, 863–869 (2001)
33. P. Matavulj, B. Timotijevic, Microw. Opt. Technol. Lett. **50**, 479–483 (2008)
34. P. Matavulj, D. Golubovic, J. Radunovic, J. Appl. Phys. **87**, 3086–3092 (2000)
35. D. Golubovic, P. Matavulj, J. Radunovic, Semicond. Sci. Technol. **15**, 950–956 (2000)
36. P. Matavulj, M. Lazovic, J.J. Radunovic, InfraredMillim. Terahertz Waves **32**, 64–78 (2011)
37. J. Petrovic, P. Matavulj, D. Qi, D.K. Chambers, S. Selmic, IEEE Photon. Technol. Lett. **20**, 348–350 (2008)
38. G. Machanovic, M. Milosevic, P. Matavulj, S. Stankovic, B. Timotijevic, P. Yang, E. Teo, M. Breese, A. Bettiol, G. Reed, Semicond. Sci. Technol. **23**, 64002–64011 (2008)
39. M. Milosevic, P. Matavulj, B. Timotijevic, G. Reed, G. Mashanovich, J. Lightwave Technol. **26**, 1840–1846 (2008)
40. M. Milosevic, P. Matavulj, P. Yang, A. Bagolini, G. Mashanovic, JOSA B **26**, 1760–1766 (2009)
41. Y. Nakano, Technologies and applications of passive optical networks (PON) [pdf] (2006). http://www.itu.int/ITU-T/worksem/ngn/200604/presentation/s6_nakano.pdf
42. G. Kramer, *Ethernet Passive Optical Network (EPON)*, vol. 9 (McGraw-Hill Professional Engineering: Columbus, OH, 2005), pp. 1–24 (0-07146-640-1)
43. G.983.1-10, Broadband optical access systems based on passive optical networks (PON) [pdf] (2005). http://www.itu.int/rec/T-REC-G.983.1-200501-I/e
44. G.984.1-7, Gigabit-capable passive optical networks (GPON) [pdf] (2008). http://www.itu.int/rec/T-REC-G.984.1/en
45. P. Green Jr., *Fiber to the Home: The New Empowerment*, vol. 2 (Wiley, Hoboken, New Jersey, USA, 2006), pp. 27–63. (978-0-471-74247-0)
46. IEEE 802.3ah Working Group, Ethernet in the first mile [pdf] (2004). http://ieee802.org/3/efm/
47. G. Kramer, G. Pesavento, IEEE Commun. Mag. **40**, 66–73 (2002)
48. M. Radivojević, P. Matavulj, J. Lightwave Technol. **27**, 4055–4062 (2009)

49. O. Haran, EPON vs. GPON, A practical comparison [html] (2005). http://www.eetimes. com/design/communications-design/40Q9354/EPON-vs-GPON-A-Practical-Comparison

50. IEEE 802.3 Ethernet Working Group, Ethernet based LANs [pdf] (2008). http://ieee802. org/3

51. M. Radivojević, P. Matavulj, in *Proceedings of the 15th Telecommunication Forum (TELFOR)* (2007), pp. 425–428

52. M. Radivojević, P.Matavulj, in *Proceedings of the 8th ETRAN*, TE4.6-1-4 (2008)

53. IEEE 802.1 Working Group, Virtual LANs [pdf] (2006). http://www.ieee802.org/1/pages/ 802.1Q.html

54. G. Kramer, B. Mukherjee, G. Pesavento, IEEE Commun. Mag. **40**, 74–80 (2002)

55. C.M. Assi, Y. Ye, S. Dixit, M.A. Ali, IEEE JSAC **21**, 1467–1477 (2003)

56. M. Radivojević, P. Matavulj, Telecommunication, 25–39 (2010)

57. M. Radivojević, P. Matavulj, in *Proceedings of the 16th Telecommunication Forum (TELFOR)* (2008), pp. 464–467

58. A.R. Dhaini, C. Assi, M. Maier, A. Shami, IEEE/OSA J. Lightwave Technol. **25**, 1659–1669 (2007)

59. R.S. Sherif, A. Hadjiantonis, G. Ellinas, C.M. Assi, M.A. Ali, J. Lightwave Technol. **22**, 2483–2497 (2004)

60. S. Blake, D. Black, M. Carlson, E. Davies, Z. Wang, W. Weiss, An architecture for differentiated services [txt] (1998). http://www.ietf.org/rfc/rfc2475.txt

61. S.I. Choi, J.D. Huh, ETRI J. **24**, 465–468 (2002)

62. G. Kramer, Y. Mukherjee Ye, S. Dixit, R. Hirth, J. Opt. Netw. **1**, 280–298 (2002)

63. A. Shami, X.X. Bai, C. Assi, N. Ghani, IEEE/OSA. J. Lightwave Technol. **23**, 1745–1753 (2005)

64. W. Willinger, M.S. Taqqu, A. Erramilli, A bibliographical guide to self-similar traffic and performance modeling for modern high speed networks (1996). http://math.bu.edu/people/ murad/pub/guide-posted.ps

65. S. Dahlfort, Comparison of 10 Gbit/s PON vs WDM-PON [pdf] (2009). http://conference. vde.com/ecoc-2009/programs/documents/sp_stefandahlfort_ng%20access.pdf

66. M. Pearson, WDM-PON: a viable alternative for next generation FTTP [pdf] (2010). http:// www.enablence.com/media/mediamanager/pdf/104-enablence-article-wdm-pon.pdf

67. N. Cheng, F. Effenberger, WDM PON: systems and technologies [pdf] (2010). http://www. ecoc2010.org/contents/attached/c20/WS_5_Cheng.pdf

68. G.694.2., Spectral grids for WDM applications: CWDM wavelength grid [pdf] (2003). http://www.itu.int/rec/T-REC-G.694.2/en

69. A. Banerjee, Y. Park, F. Clarke, H. Song, S. Yang, G. Kramer, K. Kim, B. Mukherjee, OSA J. Opt. Netw. **4**, 737–758 (2005)

70. K.H. Kwong, D. Harle, I. Adonovic, WDM PONs: next step for the first mile [pdf] (2004). http://www.comp.brad.ac.uk/het-net/HET-NETs04/CameraPapers/P47.pdf

71. M.P. McGarry, M. Maier, M. Reisslein, An evolutionary WDM upgrade for EPONs [pdf] (2005). http://mre.faculty.asu.edu/EPONupgrade.pdf

72. G.694.1., Spectral grids for WDM applications: DWDM frequency grid [pdf] (2002). http:// www.itu.int/rec/T-REC-G.694.1-200206-S/en

73. M.P. McGarry, M. Maier, M. Reisslein, IEEE Commun. Mag. **42**, 8–15 (2004)

74. M. McGarry, M. Maier, M. Reisslein, A. Keha, J. Opt. Netw. **5**, 637–654 (2006)

75. M.P. McGarry, M. Reisslein, C.J. Colbourn, M. Maier, F. Azurada, M. Scheutzow, J. Lightwave Technol. **26**, 1204–1216 (2008)

76. R.D. Feldman, E.E. Harstead, S. Jiang, T.H. Wood, M. Zirgibl, J. Lightwave Technol. **16**, 1546–1558 (1998)

77. M. Zirngibl, C.H. Joyner, L.W. Stulz, C. Dragone, H.M. Presby, I.P. Kaminow, IEEE Photon. Technol. Lett. **7**, 215–217 (1995)

78. N.J. Frigo, P.D. Magill, T.E. Darcie, P.P. Iannone, M.M. Downs, B.N. Desai, U. Koren, T. L. Koch, C. Dragone, H.M. Presby, RITENet: a passive optical network architecture based on the remote interrogation of terminal equipment [pdf] (1994). http://www.opticsinfobase. org/viewmedia.cfm?uri=OFC-1994-PD8&seq=0

79. J. Kani, M. Teshima, K. Akimoto, N. Takachio, S. Suzuki, K. Iwatsuki, M. Ishii, IEEE Commun. Mag. **41**, 43–48 (2003)

80. G. Maier, M. Martinelli, A. Pattavina, E. Salvadori, J. Lightwave Technol. **18**, 125–144 (2000)

81. J.D. Angelopoulos, E.K. Fragoulopoulos, I.S. Venieris, In *Proceedings of the 9th Mediterranean Electrotechnical Conference* (1998), vol. 2, pp. 769–773

82. J.D. Angelopoulos, N.I. Lepidas, E.K. Fragoulopoulos, I.S. Venieris, IEEE J. Select. Areas Commun. **16**, 1123–1133 (1998)

83. G. Talli, P.D. Townsend, in *Proceedings of the Optical Fiber Communication Conference* (2005), vol. 2, pp. 6–11

84. F. An, K.S. Kim, D. Gutierrez, S. Yam, E. Hu, K. Shrikhande, L.G. Kazovsky, J. Lightwave Technol. **22**, 2557–2569 (2004)

85. Y.L. Hseuh, M.S. Rogge, W.T. Shaw, L.G. Kazovsky, S. Yamamoto, IEEE Commun. Mag. **42**, 24–30 (2004)

86. F. An, D. Gutierrez, K.S. Kim, J.W. Lee, L.G. Kazovsky, IEEE Commun. Mag. **43**, 40–47 (2005)

87. H.S. Shin, D.K. Jung, J.W. Kwon, S. Hwang, J. Oh, C. Shim, J. Lightwave Technol. **23**, 187–195 (2005)

88. H.T. Lin, C.L. Chia-Lin Lai, W.R. Chang, S.J.J. Hong, Opt. Commun. Netw. **2**, 266–282 (2010)

89. K. Kwong, D. Harle, I. Andonovic, in *Proceedings of the 9th International Conference on Communications Systems (ICCS)* (2004), pp. 116–120

90. M. McGarry, M. Maier, M. Reisslein, IEEE Commun. Mag. **44**, 15–22 (2006)

91. A.R. Dhaini, C.M. Assi, A. Shami, J. Lightwave Technol. **25**, 277–286 (2007)

92. A.R. Dhaini, C. Assi, A. Shami, in *Proceedings of the IEEE Symposium on Computers and Communications (ISCC)* (2006), pp. 616–621

93. C. Xiao, B. Bing, G.K. Chang, in *Proceedings of the IEEE INFOCOM* (2005), pp. 444–454

94. Y. Tian, Q. Chang, Y. Su, Opt. Exp. **16**, 10434–10439 (2008)

95. M. Radivojević, P. Matavulj, Photonic Netw. Commun. **20**, 173–182 (2010)

96. M. Radivojević, P. Matavulj, in *Proceedings of the 17th Telecommunication Forum (TELFOR)* (2009), pp. 173–176

97. M. Radivojević, P. Matavulj, TELFOR J. **2**, 38–42 (2010)

98. M. Radivojević, P. Matavulj, in *Proceedings of the 18th Telecommunication Forum (TELFOR)* (2010), pp. 746–749

99. T. Miyazawa, H. Harai, *IEICE Trans. Commun.* E93-B, 236–246 (2010)

100. M. Radivojević, P. Matavulj, in *Proceedings of the 37th European Conference on Optical Communication (ECOC)* (2011), pp. 1–3

101. M. Radivojević, P. Matavulj, Opt. Exp. **19**, B587–B593 (2011)

102. H. Kimura, N. Iiyama, Y. Sakai, K. Kumozaki, IEICE Trans. Commun. E93-B, 246–254 (2010)

103. Cisco systems white paper, Forecast and Methodology, 2010–2015 [pdf] (2011). http:// www.cisco.com/en/US/netsol/ns827/networking_solutions_white_papers_list.html

104. SIMULINK: Simulation and model based design [html]. http://www.mathworks.com/ products/simulink

Index

A

AAL. *See* ATM adaptation layer
Access network, vii, viii, ix, x, 1–3, 5, 16, 20, 21, 30–32, 67, 69, 72–74, 92, 97–99, 114, 115, 118, 136, 144, 147, 161, 167, 168, 171, 174, 175, 187, 200, 204, 210, 217, 219
Active optical network (AON), 18
Ad hoc networks, 23
Admission control, 41, 43, 60, 61
ADSL. *See* Asymmetric DSL
ADSL2, 6, 7
ADSL2+, 6–8, 68
ADSL2++, 8
AF. *See* Assured forwarding
AF1, 50, 188–190, 212
AF2, 50, 188–190, 212
AF3, 50, 188, 189, 212
AF4, 50, 188, 189, 212
AF/BE subcycle, 136, 137, 139
Alloc-ID, 83–85, 88
AON. *See* Active optical network
APD. *See* Avalanche photodiode
APON. *See* ATM PON
Arrayed waveguide grating, 146, 147, 165–170
Assured forwarding (AF), 49–53, 129, 132, 136–140, 179, 180, 182–185, 188–191, 194–196, 198, 200, 201, 207–209, 211–215, 220, 221
Asymmetric DSL(ADSL), 5–8, 68
ATM
    adaptation layer, 77
    adaptation method, 79, 86
    cell, 75–77, 86, 97
    cell tax, 77
    PON, 68, 69, 89, 97, 173
Attenuation, 12, 82, 149
Auto-discovery, 102
Avalanche photodiode (APD), 149

AWG. *See* Arrayed waveguide grating

B

BA. *See* Behavior aggregate
Band splitter(BS), 146, 147
Bandwidth allocation, 36, 57, 83–85, 88, 102, 109, 118, 120, 121, 125, 127, 130, 131, 133, 138, 141, 143, 152, 159, 161, 171–173, 180, 182, 185, 186, 190, 196, 206, 221
    algorithm, ix, 94, 130, 133, 138, 168, 178
    centralized, 119
    distributed, 119
    dynamic, 75, 77, 83, 84, 107, 110, 120, 128, 145, 152, 166, 172, 176, 191, 207, 219
    fixed, 72, 77, 122
    grant sizing, 159, 161, 163, 178
    scheme. *See* Grant sizing,bandwidth allocation algorithm
    static *See* Fixed
    unused, 135, 182, 183, 187, 196
    upstream, 73, 77, 80, 88, 94, 109, 135, 159, 170, 178, 180, 182, 185–187
Bandwidth assigned mode of operation. *See* Normal mode of operation
Behavior aggregate (BA), 48, 54
BER. *See* Bit error rate
Best-effort service, 40
Best-effort traffic (BE), 49, 130, 132, 136, 138, 140, 168, 179, 184, 185, 189, 198, 207, 211, 221
BIP. *See* Bit interleaved parity
Bit error rate, 78
128-bit bitmap, 156, 158
Bit interleaved parity, 85
BPL. *See* Broadband over power line
Broadband, vii, 1–3, 6, 11, 14, 16, 22, 24, 26, 28, 30, 32, 75

© Academic Mind and Springer International Publishing AG 2017
M. Radivojević and P. Matavulj, *The Emerging WDM EPON*,
DOI 10.1007/978-3-319-54224-9

Printed in the United States
By Bookmasters